THE SACRED MUSHROOMS
OF MEXICO

Assorted Texts

Edit
Brian I

T0127898

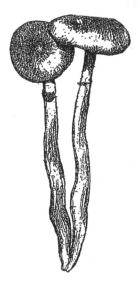

University Press of America,® Inc.
Lanham · Boulder · New York · Toronto · Plymouth, UK

Copyright © 2007 by
University Press of America,® Inc.
4501 Forbes Boulevard
Suite 200
Lanham, Maryland 20706
UPA Acquisitions Department (301) 459-3366

Estover Road
Plymouth PL6 7PY
United Kingdom

Library of Congress Control Number: 2006930458
ISBN-13: 978-0-7618-3582-0 (paperback : alk. paper)
ISBN-10: 0-7618-3582-2 (paperback : alk. paper)

To the memory of
friend and colleague

Margaret S. Houston

who helped make this
book possible

Contents

Preface

Another book on the sacred mushrooms—why? In recent years, the number of titles offered on this subject seems to have grown. The groundbreaking works that established this field of inquiry, the studies of the late R. Gordon Wasson and others, are now celebrated classics of scientific discovery and multi-disciplinary study. Why the present volume?

A considerable general interest in this subject has long been evident to me from inquiries and correspondence I receive in private, and by the number of new titles addressing it that have been published in recent years. I can attest to the fascination exerted by the sacred mushrooms of Mexico from my own point of view, both professionally and personally, having grown up in the era during which their story unfolded. As a mycologist with a long standing interest in ethnobotany and a graduate degree in anthropology, I have perhaps had more requests to address audiences on the subject of ethnomycology than any other. The occasions when I have been able to give such presentations have been enjoyably memorable for me. For the honor of such invitations I owe an acknowledgment to the interest and kindness of groups such as the Missouri Mycological Society, the Gulf States Mycological Society, the Asheville Mushroom Club, the Minnesota Mycological Society, the New Jersey Mycological Association, the Mycological Association of Washington, and the North American Mycological Association.

Ethnomycology is a subject that has commanded my interest for years. In studying toward my Masters in Anthropology from Western Michigan University, my focus was ethnobotany, especially as pertains to indigenous uses of hallucinogenic plants. My thesis was titled *Peyote and Peyotism*, and besides studying sources specifically addressing that subject, I read widely to help secure a basis for comparative understanding as broad and sound as possible. This was the context in which I began delving into the matter of the sacred mushrooms, which had been compared with peyote from almost the moment of Western contact with native Mexican culture.

In the course of my ethnomycological studies, I noticed recurring references in various articles and books to certain works in other languages, especially Spanish. Judging by their repeated citations in relevant literature, I realized these sources were likely to contain significant information and discussion. As an example, consider the following quote from Gastón Guzmán, internationally noted Mexican mycologist:

> Among other 'post-Wasson' works on the ethnomycology of the sacred mushrooms, Hoogshagen (1959), Miller (1966), Ravicz (1960) and Lipp (1971) presented interesting findings from their studies in the State of Oaxaca (Guzmán 1990, 100).

The first and last articles referenced above are definitive studies representing important contributions to the English literature on the subject. But the works by Miller and Ravicz are in Spanish, and this posed a problem for my wish to read them due to my lack of fluency in that language. Nonetheless, seeing these Spanish works cited by noted authorities in the same breath as other sources I had found informative greatly kindled my interest. Based on library research, I determined that there were perhaps, in particular, five literature sources in Spanish that stood out as items worthy of serious attention and study: Ravicz (1960[1961]); Miller (1966); Benítez (1970); Reyes G. (1970), and Escalante and López G. (1972).

Assisted in my literature searches by courteous staff of Inter Library Loan (ILL) offices at several colleges, I have generally been able to acquire the majority of sources relevant for my work, including most Spanish sources. Over the years, I have gradually worked up English translations of these, at first with assistance strictly from flesh and blood resources, and more recently with the further aid of computer translation engines. Now that I am able to peruse and study these texts in English I have found that, individually and collectively, they do indeed offer a fresh and fascinating vista. Together with the transcript of *One Step Beyond* in the last chapter, the texts offered herein are all vintage pieces dating from 1960 to 1972. They are core works from the classic era of ethnomycological discovery in Mesoamerica, which began when the Wassons entered the field with their first Mexican expedition in 1953, and continued until R. G. Wasson's death in 1986. As such, these works deserve their place alongside the other significant texts on this subject available in English.

Of the five Spanish literature sources of interest I had identified, I found that four were readily obtained through college Inter Library Loan offices. But one seemingly defied my efforts: *Hongos sagrados de los matlatzincas* by Escalante and López (1972). Citations to this source can be found in such authoritative works as Ott and Bigwood (1978), Wasson (1980), and Guzmán (1990). Over the course of my studies, I made repeated efforts to obtain this article through library offices spanning seven colleges and universities where I was successively stationed from 1981 to the present. In each case the results were negative. Apparently, this was a rare paper, not to be found among the collections of any library within reach of ILL.

Having exhausted such avenues of inquiry, I finally ventured to contact the world's leading authority in the genus *Psilocybe*, Dr. Gastón Guzmán, to inquire whether he could assist my efforts. He happened to have a copy of this article, and to my immense gratitude he went to the trouble of electronically sending me digital photographs of the pages, one by one. This is how I finally acquired the text underlying my English translation of Escalante and López, as offered in the present volume. I am deeply indebted to Dr. Guzmán for his courteous and collegial help, and it is with pleasure I take the opportunity here to express my appreciation to him. Thanks to his generosity, we English readers may now better regard the whole crown of these Spanish works, with all five of its jewels properly in place, at least as I have them translated.

It has taken years for me to develop and finalize these translations to a point where, although by no means perfect, they can be presented for general consumption. One might almost regard the combined works of Wasson and his collaborators in English and French (since some of Wasson's most important collaborative works were in the latter language), as the veritable Old and New Testaments of ethnomycology. By this analogy, the translated pieces offered herein could be thought of as extra books, English versions of which have been lacking until now. In a sense they already are part of the canon, based on the recognition they have received from citations in other authoritative literature.

My efforts to get these Spanish sources translated thus originated purely out of a simple wish to read them for myself, to improve my own understanding of the subject matter. Once I had achieved this goal, I found that they were indeed of considerable interest and well worth the trouble to which I had gone. I thus conceived a further desire at this point to make them available to other readers, and present them in homage to the general interest in this subject, which has been generous to me. When I first learned of the ethnomycological writings of Robert Ravicz, Luis Reyes G., Walter Miller, Fernando Benítez, and Roberto Escalante H. and Antonio López G., I lamented that they were not available in English. That no one had already translated and published them for students such as myself, eager to peruse them, was a frustrating circumstance to me. On this basis, I consider there may likewise be others who will be similarly interested in reading these works now that English versions are available.

The transcript of *One Step Beyond: The Sacred Mushroom* fits well into the analogy of a literary canon, albeit as an apocryphal text, perhaps even heretical. It has never before been considered in any authoritative ethnomycological account, and is no less interesting for it. This show was certainly one-of-a-kind in the history of broadcast television, and there are many who have never had a chance to view it. But those who have tend to remember it. There are not many television programs one can cite, documentary or otherwise, in which the affable weekly host traveled to Mexico to film sacred mushroom ceremonies, and then brought back samples to consume on camera, withstanding the onslaught of their effects for the viewing pleasure of the audience at home in the American living room of January, 1961, all in the context of testing for enhanced psychic ability. In the history of ethnomycology, the *One Step Beyond* chapter stands as a unique kind of chronicle, distinctly different from the translated Spanish literature sources. As such, it offers a striking reflection of the popular impact of the widespread publicity surrounding the sacred mushrooms in that halcyon era. Relations between the show and the history of ethnomycological studies behind the scenes also prove interesting, as revealed herein. It is a pleasure to offer this chapter side by side with the other pieces newly translated here.

I thus submit these texts for the approval of readers, as delicacies to tempt their interest, and tokens of my appreciation for it. I also offer them as a modest tribute to the dedicated researchers who have contributed so much to our knowledge and understanding of the sacred mushrooms of Mexico, so that people such as the rest of us may learn from their achievements. Among these contributors R. G. and V. P. Wasson and their various associates figure, far and away, first and foremost. In this connection, the greatest credit and gratitude of all must go to Luis Reyes G., Roberto Escalante and Antonio López G., Walter Miller, Robert Ravicz, and Fernando Benítez, the authors whose works inspired the idea for this volume, and whose substance and meaning hopefully come across in the translations presented herein. This book thus goes out as a salute to the value and merit of their contributions, ripe as they are for appreciation by the English speaking world, even overdue.

The scholarly foundations of ethnomycology have an almost labyrinthine structure, intersecting many disparate fields of expertise deeply and tortuously. Indeed, one reason the ethnomycological paradigm has been so compelling lies precisely in the way separate, highly divergent disciplines are seamlessly integrated under its recognition of mycophobic and mycophilic culture patterns, and its explanation for the likely origins of this cross-cultural opposition in humanity's presumed encounters with psychoactive fungi in prehistory. The sacred mushroom complex of Mesoamerica is of key relevance for this paradigm. The connections between shamanistic practices in the New World and Old, and the ancient roots of religions in the modern world, have perhaps never been as extensively and startlingly illuminated as they have in light of the discoveries and analyses of the Wassons and their collaborators.

But by this same token, the number of different disciplines drawn upon by ethnomycology makes it a challenge for study, one of considerable proportions. In working with the texts in the present volume, technical aspects of mycology, of anthropology, and of other subjects have proven to be crucial. In such literature, the ordinary difficulties of translation are compounded by the technical use of certain terms from various disciplines which also have general, less specific meanings outside those disciplines, such as type (see p. 159). Or again, whether a word such as "cap" refers to a garment for the head, or a part of the fruiting body of a fungus is already a critical distinction. But it is even more so when both meanings may be germane, as in studies that draw equally upon cultural anthropology and mycology. In translating ethnomycological sources, one deals with the idioms not only of language but also of culture. Faced with many vagaries in the course of this work, I have been greatly advantaged to have colleagues in several fields upon whom I have been able to rely for expert comment in interpreting the Spanish sources presented herein. With its inherent complexity, ethnomycology has proven to be a triumph of interdisciplinary scholarship, but also a rigorous, multi-faceted path of study.

With these considerations in mind, a few words are in order about how this book has been composed. I assume that some readers will have more study in anthropology than mycology, or vice-versa, while yet other general readers will have little or no background in either. In the portions of text composed by myself, I have therefore tried to render the subject matter broadly accessible by providing contextual information, whether from fungal biology or cross-cultural study, to help clarify it. For each chapter I have included an opening discussion to introduce the text that forms its core (plus, in the case of the *One Step Beyond* chapter, an Appendix with additional relevant information). To help further enhance the materials presented, I have also inserted editorial notations here and there, in one or both of two forms; (1) brief remarks in parentheses, containing the abbreviation Ed.; and (2) in the Benítez chapter, "Translator's notes" marked in the text by a superscripted letter of the alphabet, rather than numeral. I have used letters instead of numerals for this purpose because numerals are already in use by the original author for notes in the Spanish text, and I have preserved them in that form for my English rendition. Using letters to mark Translator's notes has allowed me to add comments that may help further clarify the readings, while avoiding any confusion that might otherwise arise between the editorial voice and that of the original author. Indeed, numeric superscripting already does double duty in some of the sources herein, in the original author's notes, and in linguistic notation of native terms, for example $\check{c}o^4ta^4ni^4\check{c}e^4$, or other words in tonal languages such as Mazatec. My alphabetic system limits me to twenty-six Translator's notes, but this allowance has proven to be quite sufficient. My added notes follow the original author's notes, at the end of the chapter in which they appear.

Speaking of $\check{c}o^4ta^4ni^4\check{c}e^4$, a general comment is perhaps in order regarding transliteration of Indian terms, and the notation one encounters in these readings. I have not studied linguistic orthography, but it is clear that a number of different systems are in use from one source to another. These range from more detailed and technical styles, such as that employed in the example given above (i.e., including non-English letters with diacritical marks, and superscript numerals in the case of tonal languages, etc.) to less so (no numerals, and purely English alphabet). Differences between words or names observed in different sources may in some cases simply be artifacts of differing styles of notation. As a result it has not been possible for me to clarify or correlate every reference to indigenous names, unfortunately. Some untidiness remains.

Consider the following example, concerning Mazatec names for *Psilocybe cubensis*, one of the more familiar and widely distributed psilocybin mushroom species, and one recognized as sacred in Mexico. To the Mazatec Indians, it is known by a name in their language which means "sacred mushroom of cow manure." This name has been variously rendered in literature, such as *'nti¹ si³tho³ y'e⁴le⁴ nta⁴ ha⁴* (Heim and Wasson 1958, 53), and *di-shi-thó-le-rra-ja* (Ott and Bigwood 1978, 108). Singer (1958), gives it as *di-shi-tjo-le-ra-ja*, and adds in a footnote:

G. Cowan spells this 'nti¹-xi³-tjo³-le⁴-nčha⁴-ja⁴, pronounced *ndee¹-shree³-t(h)oe³-lay⁴-njra⁴-ha⁴* (meaning as above, dear little thing that comes out pertaining to the steer or ox). The pronunciation of *čh* is described as a sound similar to English *j* (in judge) and *r* (in run) if pronounced simultaneously (1958, 251ftn).

In further connection with this matter of linguistic notation (should the preceding not suffice) I also cite from Singer (1958) a comment which

> Mr. George M. Cowan, a student of the Mazateco language, was kind enough to offer: "We use ... a practical orthography ... This is simply a phonetic alphabet adapted as far as possible to the Standard Spanish alphabet. Tone is significant in Mazateco; the numbers ¹ ² ³ ⁴ represent relative pitch of the syllables; ¹ represents the highest, ⁴ the lowest. (The symbol) ' (i.e., an apostrophe—Ed.) cannot be transcribed in English orthography and is not heard by the untrained ear; it is sort of a glottal catch preceding the following sound." The correct phonetic spelling of *di-nizé* (Ed.—a Mazatec name for *Psilocybe mexicana* reported by Singer) is, according to Mr. Cowan, 'nti¹-ni⁴-se³; it is pronounced ndee¹ nee⁴ say³... (Singer 1958, 250-251).

I hope these quoted passages and the obscurities of language and pronunciation they reflect help convince the reader not to be too daunted by these formidable-looking transliterations, nor to invest too heavily in specific details of their forms, unless perhaps one is specifically interested in linguistics. I have been careful to preserve these notations faithfully as they appear in their original sources. But as this example of the Mazatec name for *Psilocybe cubensis* illustrates, they need not have too fine a point put on them, and I encourage the general reader to view them in that light. The scientific binomials for various species, and the changes of nomenclature they undergo, seem easy by comparison.

Regarding the Spanish-English translations, there have been many questions of how to render them, as to both the words and meanings. In certain passages, I have reproduced in parentheses, following the English translation, the Spanish words or phrases used in the primary source. Sometimes it is more important (*más importante*) to know what the original word or phrase was, than how one has translated it, which after all is secondary, partly a matter of decision and choice. On the whole, I have attempted to strike a balance between imperatives of English style and the syntactical structures of the original Spanish, which have their own characteristic rhythm and sequencing generally different from that of English. I have also opted to preserve stylistic differences between the various sources presented herein, rather than standardize them for the sake of the book format, so as to better enable comparative study between the translations offered herein and the original sources upon which they are based.

For word meanings, especially obscure or difficult ones, a particularly helpful source has been *A New Pronouncing Dictionary of the Spanish and English Languages*, (1957) compiled by Mariano Velásquez de la Cadena et al. (using a copy loaned to me by Dr. Margaret Houston). In Mexico, the vocabulary relating to mushrooms involves a combination of Spanish and indigenous terminologies. There are perhaps only two terms in Castilian Spanish for mushrooms, *hongo* and *seta*. The former is cognate with the word fungus, and the latter is a general term for various kinds of mushrooms; particularly the field mushroom *Agaricus campestris*, according to the *New Pronouncing Dictionary*. As an interesting aside, the Wassons note that in Castilian tradition only two types of mushroom are eaten, adding:

> The poverty of the Spanish language when it comes to mushrooms occasionally works a hardship on Spanish writers. In the Spanish encyclopedia the authors of the articles on *hongos* must resort to Catalan and Basque words to piece out the poor Castilian vocabulary (Wasson and Wasson 1957, 339).

Speaking of language, I am deeply indebted to several people who have worked with me on the translations contained here. Their contributions have been invaluable, and this book would not have been possible without their help. Indeed, for three of the translation pieces presented, the lion's share of the work has been theirs. I am pleased to credit them fully here for all they have done and to thank them for their assistance. But I must claim responsibility for any errors or rough edges in the translations, since I am the one who has exercised sole discretion over the final form in which they appear. With this understanding I thank Mr. Richard de Meÿ for his fine translation work, at the University of Florida in 1991, on the article by Miller on the Mixe *tonalamatl* (or "Book of Days"). His translation appears herein virtually intact. For the account by Reyes G. concerning the *nanakatsitsen* ("the little mushroom people"), assistance with translation was provided in 1982 by anonymous staff at the Language Center at Western Michigan University. I would add that although the Center's service fee was my own expense, the University provided me with stipends in the form of a Graduate Fellowship and Assistantships, through the Anthropology Department. For the Ravicz chapter, and various odd questions arising from several others, my work was very kindly and expertly assisted by Dr. Margaret Houston, whose understanding of the cross-cultural aspects of Mesoamerica and nuances of language there is unparalleled in my experience.

However poor the Castilian vocabulary for fungi, the reader will find in these texts a rich and diverting blend of fact and traditional folklore, full of colorful and exotic elements. Examples would include the two references in the preceding paragraph to Indian terms and concepts, one (the *tonalamatl*) a sort of native almanac or calendar book, and the other (the *nanakatsitsen*) less tangible, almost a counterpart to elves or leprechauns, but perhaps even smaller, part of the mythos and lore of the mushroom in Mexico. Besides the *nanakatsitsen*, one will also make nodding acquaintance in these pages with such obscure little beings as *duendes, chaneques, enanitos*, and *chamacos* y *chamacas*, all of them miniature mythic personifications from regional native lore. Another class of beings, of even greater supernatural power, is the *dueños*, mystical owners or masters of various elements, resources, or locales, on account of which they often command some deference. There are also native deities of greater or lesser renown including Tláloc, Coatlicue, and others. At some point one may ponder an almost Tolkienesque quality about these figures, except that they originate from the real world of indigenous Mexican cultures, rather than the fiction of English literary fantasy.

The empirically based elements encountered herein are no less interesting. Various types of indigenous practitioners and words for such appear, each apparently connoting something slightly different from the others, with the differences themselves seeming to permute from one region to another. These include *curandero* or *curandera*; *brujo* or *bruja*; *hechicero* or *hechicera*; and *sabio* or *sabia*, the latter corresponding to the Mazatec *čo⁴ta⁴ni⁴če⁴*. The Wassons note that the term *curandero* expresses respect, whereas *brujo* is more pejorative (1957, 250ftn.). Benítez, in his feature in this volume, asks the famous Mazatec *curandera* María Sabina about this distinction. She ought to know since, as she explains, she was married to a *brujo* at one time. (A glossary to some of this vocabulary is provided toward the end of this volume.) For English analogies, we might consider the shaded differences of meaning, especially in terms of implied value judgments, between designations such as native healer, medicine man, and witch doctor.

Indigenous place names of decidedly authentic sound come in an unending stream in these pages, starting with Huautla de Jiménez in the Sierra Mazateca, and going on from there. Most of the locations discussed, including the latter, are in Oaxaca. Likewise, one is introduced to, if not already acquainted with, various groups of Indians with names perhaps not as familiar to most English readers as those from north of the border, names such as the Chatino, the Zapotec, Mixtec, Mixe, Mazatec, Matlatzinca, etc. Elements of everyday life such as the attire, the *huipil* or traditional embroidered blouse of the women, the *sarape* or shawl; or cooking implements such as the *mano* and *metate* used for grinding; or *aguardiente*, a sort of home distillate or moonshine derived from sugar cane, paint the scenes with vivid detail. With the combination of all these elements, studies on the sacred mushroom provide a particular kind of window on Mesoamerican culture, and are of ethnological interest for this reason among many others.

I have many colleagues senior and junior to thank as well as friends and family, the latter particularly including my surviving siblings Diane and Al Akers. I also owe a hearty thanks to all those who have joined me in fungal and ethnomycological interests over the years, particularly members and officials of various mushroom clubs I have had the pleasure and good fortune of getting to know a bit, such as Gene Yetter, Jay Justice, Ken and Laree Gilberg, Maxine Stone, Barkha Bullin McDermith, Andrea Vadner, Jack, Marty and Lee Toll, David Sacks, Barb O'Brien, Don Dill, Dean Abel, Linda Mueller, Claudia Joyce, Mary Brent, David Yates, Patrick Lennon, Ellen Wheeler, Brad Bomanz, Bill May, Leland Von Behren, Ron Spinoza, Robert Fulgency, Anna Gerenday, Carol Dreiling, Pete Whelihan, Renate Rikkers, David and Patricia Lewis, Kurt Rottweiler, Susan Hopkins, Glen and Ania Boyd, Alan Orovitz, Rhoda Roper, and many others too numerous to name, for their extraordinary generosity and hospitality as well as supportive shared interest in fungi and ethnomycological studies. Such gracious people, so appreciative of the allure of wild mushrooms, make for highly enjoyable company in fungal pursuits.

Speaking of shared interest in fungi, I would also like to extend a salute of appreciation to my graduate mentors, particularly the distinguished mycologists Dr. James Kimbrough of the University of Florida, and Dr. Walter Sundberg of Southern Illinois University at Carbondale. The present volume could never have been possible without the cordial and professional assistance of these gentlemen, who patiently oversaw my training. I have also benefited from and enjoyed my professional associations with colleagues in the field of mycology such as Dr. Clark Ovrebo, Sherri Angels, Dr. Gerald Benny, Dr. Lorelei Norvell, Dr. Coleman McClenegham, Dr. Alan Bessette, Dr. Barrie Overton, to name a few. The same must be said for various professors in related sciences who have also kindly assisted my efforts in a multitude of ways, including Dr. Mike Korth, Dr. David Lightfoot, Dr. John Bozzola, Dr. Lawrence Matten, Dr. Don Tindall, Dr. Walter Schmid, Dr. Terry Lucansky, Dr. Walter Judd, Dr. William Stern, Dr. Donald Ugent, Dr. Allen Dotson, and Dr. Aristotle Pappelis. I have also benefited from the supportive interest and assistance of professors outside the natural sciences, including Professors William Rolland and his wife Joy, Bill Alexander and his wife Kay, Carl Walters, Ron Bayes, and with a special thanks to Huston Smith. Many helpful interactions and inspirations over a span of years underlie the present volume. I have enjoyed and appreciated the interest of Dr. and Mrs. Arthur Welden. Dr. Kimbrough's wife Jane has also been a generous, tremendously kind and supportive force over the years, and I would not miss this moment to offer her my thanks. My studies beyond the boundaries of science have been encouraged by colleagues at the Metanexus Institute, particularly Barbara Bole and Dr. David Hufford, and elsewhere such as Lou Ann Trost, formerly editor of the Bulletin of the Center for Theology and Natural Sciences (CTNS). I also appreciate the interest and support of individuals who have participated in activities funded by the Metanexus Institute in which I have been involved, such as Alan and Beth Hui, Calvin Dittrick and his wife Adele, Jim and Gail Lockamy, Willie and Ruby Thomas, Jack and Evelyn Hanna, Catherine Neylans, and many others.

I must also thank the officials and curators of museums and herbaria, and staff of archives, institutes and libraries who have responded to my endless requests for assistance, particularly Héctor Toledano, Publications Director of the National Institute of Anthropology and History (INAH) in Mexico, Lisa DeCesar of the Botany Libraries at Harvard University, Claire Margerie of the National Museum of Natural History in Paris, Paul Hoogstraten of Pi Publications in the Netherlands, Karen Dakin of the Instituto de Investigaciones Filológicas, Mariana Palerm of Ediciones Era, and ace Inter Library Loan specialist Mary MacDonald, whose sterling efforts on my behalf have repeatedly amazed me with their results. Indeed, I have come to regard Mary as my veritable High Priestess of ILL. In addition, I would like to express a deep and special thanks to Dr. George Melton, Professor of History. It was he who gave me, as a present from his collection of *LIFE* magazine back issues, the copy of

"Seeking the Magic Mushroom." Thank you George, I will never forget your kindness.

Friends of long acquaintance from schooldays past such as Ken Greene and Scott Anderson, Jim McGahey, Mark Barnes, and others of more recent and scientific professonal vintage such as fellow biologists Nancy Wilson and John Moeller, have been fountains of inspiration and deeply appreciated camaraderie, and deserve a special thanks. Fellow graduate students from days gone by such as Steve Morton, Scott Franklin, Jonathon Newman, Jody Shimp, Mark Basinger and Erika Grimm Choberka and her husband David, Paul Monaghan, Nick Ciccotosto and his wife Carol, plus a score or two more also merit words of appreciation for their friendship and good company. Along similar lines I would like to also thank Donovan Barbara, Vince Whitmore, Hal Pulfer, Jeff Hessler, and Tim and Kathy Morris. I also appreciate technical assistance I have received from friends such as Diane Hanke, and Chris Siefken whose wife Jessie likewise provided helpful comments about various translation points. In addition, I thank Catalina Ramirez for her capable assistance, and my "legendary" associates Jeffrey Dahlgren and Julia Blue. A word of thanks also goes to Steve Lapping for his help, and to his wife Rebecca. In addition, I especially thank John Kenneth Muir for his shared interest and extraordinarily generous assistance concerning my research on *One Step Beyond: The Sacred Mushroom*.

I also wish to acknowledge some of my recent junior colleagues along the way such as James Hocker, Jaysen Walker, Viktoria Trebeleva, Elizabeth Camp, Sylvia Outley, Nancy Hoffman, Marcus Womack, Neville Beamer, Mark Hawkins, Randy Hughes, Bridgid Raleigh, Patrick O'Donnell, Kimberly Rei Simon, Molli "Rocket" Leidel, Jennifer Johnson, Lanisha Humbert, Matt Dowling, Gilbert Abraham, Noelle Braxton, Andrea "Danger" Eaton, Erin Hughes, John Roth, Ryan Campbell, Martha Mabry, Liz Pisarczyk, Janelle Munter, Sammi Holsinger and her mother Gail Anderson. Other junior scholars of recent acquaintance also merit recognition, such as Ian Wallace, James White, Seth Wells, soon-to-be newlyweds Ben Morissey and Audrey Rolfe, Rosanna Borders, Sarita Houston, Kristen Hefner, Amanda Fletcher, Lloyd Blevins, Ashleigh Grimes, Renee Gubert, Jenny Smith, "Mustang" Andrea Johnston, Dade Webster, Patrick Kretzchmar, Andy Hendrix, John Williams, Suzanne Crowe, Brandy Gilbride, Chuck Bond, Mary Beth Reynolds, Adryn Henard, Meredith Fish, Quintin Carter, Faye Warren, Patrick Choogog, Jacob Reading, Tiffany Blanchard, Kalifa Bojang, Sarah Rhymer, Mike Genest, Shelley Shoe, Ryan Curley, Emily Goulet, Jesus Gonzalez, Katie Baringer, Aaron Gatten, Christian Yingling, T. Jordan Johnson, Alexx MacLennan, Kelly Rothlisberger, Ashley Colvin, Mike Squillante and many others. To all of you I offer acknowledgment, high regards and thanks. And I hope you enjoy this book.

In many spheres of life and living, it seems the truth is often stranger than fiction. The fungi in general would stand as a case in point, I think. But the story of hallucinogenic fungi and their human relations across culture, through history and in the prehistoric past, strikes me as a most considerable example of this principle. The more ethnomycological study has revealed in the last half century—and the 50 year anniversary of the historic May 13, 1957 issue of *LIFE* magazine is fast approaching—the more amazing the story has seemed, with discovery upon surprising discovery in succession, often fraught with irony. At the interdisciplinary frontiers of this field, I have sometimes found separating fact from fancy to arrive at a sound perspective to be no simple matter. I hope the present offering can contribute in some small way to the attainment of better understanding of such a complex subject.

Brian P. Akers
Laurinburg, North Carolina
May 20, 2006

Permission to publish English translation from Heim, R., 1967, *Nouvelles investigations sur les champignons halucinogènes* granted by Le Service des Publications Scientifiques. Archives du Muséum national d'Histoire naturelle 7è série, tome IX. Muséum national d'Histoire naturelle, Paris, 229 p. Extracts: pages 160-161, and 219. © Publications Scientifiques du Muséum national d'Histoire naturelle, Paris.

Permission to republish quotes from Wasson, V. P., and R. G. Wasson, 1957, *Mushrooms, Russia and History*, Vol. 2, granted by Tina and R. Gordon Wasson Ethnomycological Collection Archives, Harvard University, Cambridge, Massachusetts, USA.

Permission to publish English translation of Reyes G., L., 1970, *Una relación sobre los hongos alucinantes*, granted by *Tlalocan*, Seminario de Lenguas Indígenas, Instituto de Investigaciones Filológicas of the Universidad Nacional Autónoma de México.

Permission to publish English translation of Benítez, F., 1970, *Los hongos alucinantes*, granted by Ediciones Era, México D. F.

Chapter One
Introduction

The present volume contains six texts—five translations, and a transcript—which will be of special notice for those interested in the subject of psilocybin mushrooms and their traditional shamanistic use in Mesoamerica. The ritualized use of fungal hallucinogens is not without its analogies, but in Mesoamerica its manifestations have been striking and unique, with impressive local variation and underlying unity suggesting a considerable antiquity.

This intriguing and broadly based cultural manifestation may be generally referred to as *nanacatism*, from *nanacatl,* a term for mushroom in Nahuatl, the language of the Aztecs and related groups collectively known as the Nahua (Wasson and Wasson 1957, 219). As noted by Schultes (1940) a Hispanicized form of the latter word, *nanacate*, persists to this day in Mexico. Some of the earliest sources indicate the term *teonanácatl,* in which the prefix *teo-* means wondrous or sacred, was one of various designations used by Nahua peoples for psilocybin mushrooms (awkwardly termed 'psilocybian' in some popularized references). Word pairings such as sacred mushroom or magic mushroom are thus reasonable English approximations for this particular native term, one of several among the native lexicons referring to these fungi.

To date perhaps twenty or so species of mushrooms, mostly in the genus *Psilocybe*, have been reportedly recognized as sacred and utilized as such among various contemporary native peoples of Mexico. Cultures in which some form of nanacatism has been documented in modern times include the Nahua, the Chatino, the Chinantec, the Matlatzinca, the Mazatec (or Huautec), the Mixe (or Mije), the Mixtec, and the Zapotec. The Mesoamerican landscape is extensively dissected by mountains and valleys, and its habitation reflects this fact. Isolation and diversification have prevailed over time, yielding a rich mosaic of distinctive local tradition, and considerable biological speciation. Some of the mushroom species used in rituals are widely distributed, but others are known

1

from only one or a few localities, apparently having much more restricted ranges.

Origins of Ethnomycology

Historically, the first records of the native use of sacred mushrooms were accounts by a few of the Spanish chroniclers in sixteenth century Mexico, such as Sahagún and Hernández (Schultes 1940). By 1957, Wasson and Wasson had identified ten such early authors; and in years to come that number grew along with interest in the subject (Wasson 1980). These early historic references to the sacred mushrooms were brief, but indicated they were called by names such as *teyhuinti*, which means sacred intoxicant, and, more famously, *teonanácatl*.

Nanacatism also seems to have been present at one time in Guatemala and neighboring regions. This is evidenced by archeological artifacts, especially the so-called "mushroom stones," which date from the highland Mayan pre-Classic and Classic civilizations, roughly 1000 BC to 500 AD. There are in addition some suggestive ceramic artifacts from Colima, Nayarit and Jalisco in northwestern Mexico (Furst 1974). Indeed, the cultural foundation for nanacatism ultimately encompasses the entire neotropics, and the wide variety of hallucinogenic plants used in Mesoamerica and South America (Schultes 1972). The idea of ritual use of such agents was probably already well in place culturally when the New World first became inhabited. It is likely that a sustained, deliberate search for plants and fungi of interest for such purposes has occurred, considering the sheer number and variety of species used in the New World narcotic complex as La Barre (1972) has termed it. The shamanistic use of hallucinogenic mushrooms would thus seem to have a substantial antiquity with cross-cultural roots in Mesoamerica based on total evidence, especially as amassed and illuminated by interdisciplinary investigations over the past century.

The modern world's fascination with the sacred mushrooms, and our present knowledge about them, can be traced in large part to the research of R. Gordon Wasson (1898-1986) and his various collaborators. First and foremost among these was his wife Valentina Pavlovna Wasson. Sadly, she died in 1958, soon after their Mexican studies had just begun to bear fruit. Before the Wassons' involvement in the subject, nothing was known of the chemistry or pharmacology of these mushrooms, nor was there any significant understanding of the species that figured in their traditional use. Indeed, as we now realize, most of the species regarded as sacred were completely unknown to fungal specialists, and were first described and named in connection with field expeditions Wasson organized, with careful follow through afterward in the laboratory. He and his associates brought the subject of nanacatism, which at mid-century was languishing in a state of relative obscurity, almost unheard of, to the attention of the general public, as well as specialists in many divergent fields of study upon which it has bearing. Wasson did not discover *per se* that there was such a thing as hallucinogenic mushrooms, nor that they were deliberately used for their effects by certain indigenous peoples in shamanistic rituals. Instead, the core of his contribution was a startling realization of the sheer impact fungi have had upon human history and culture in general—which turns out to be extensive, and quite a bit greater than anyone might have previously supposed. It is interesting that Allegro (1970), working independently of the Wassons with evidence adduced from study of ancient languages, came to conclusions which agree with this larger perspective, although many of its specifics are not supported by Wasson's research.

This impact of fungi upon humanity is evident in comparisons across the boundaries of culture and language. The English speaking world, with its relative lack of traditional folk names for the various types of wild fungi, and its fanciful, distance-keeping expression *toadstool*, is one example. This peculiar lack of common names for most wild mushrooms, even the more familiar kinds, becomes conspicuous only by contrast with other languages and traditions. As native English speakers we may never stop to think about it, but this deficit of fungal folk names is remarkable considering that so many insects, wildflowers, birds, trees, and even varmints, all have theirs. Viewed in comparative perspective, this traditional non-christening of wild mushrooms expresses a distinct cultural pattern. The Wassons (1957) described this as mycophobia, and it contrasts sharply from mycophilia, such as can be seen in the French, or eastern European cultures, for example. This helps explain why names from other languages must be imported, such as *shiitake, chanterelle,* or *morel,* or even coined anew (*portabello*), as more and different kinds of produce mushrooms become commercially popular at our local fruit and vegetable stands. The English mycophobic orientation is also richly expressed by the frequent recourse to fungi in poetry and prose as icons of repugnance and disgust, reminders of death and decay and everything rotten in the world. The Wassons (1957) cite various examples in this regard (although I find no references to a conspicuous and celebrated example, the noted, eccentric, early twentieth century American writer of weird fiction, H. P. Lovecraft).

This distinct divergence between culturally patterned aversion to versus celebration of wild mushrooms, the fungal citizens of nature's realm, in itself reflects the antiquity and power of a primal human encounter with fungi. In the course of their research exploring this cross-cultural schism, the Wassons founded a field of study known as ethnomycology, a subdiscipline or sister discipline of ethnobotany. Their work required expert technical advisement from various fields, and thus they made it a point to always seek out the assistance of accomplished and recognized authorities in these fields. In payoff, discoveries made by the Wassons and their expert collaborators have shed light upon subjects as widely separated as anthropology, mycology, and pharmacology. Findings by the Wasson team have opened many doors to understanding, a clear indication of their scope and importance.

Based on compelling evidence, V. P. and R. G. Wasson proposed in 1957 that the impact of fungi upon history and culture, and the explanation for the divergence and global dispersion of mycophilic and mycophobic cultural patterns, lay in the profoundly mind-altering effects of substances found in certain species, and the native interpretations placed upon those effects as directly experienced. For the Wassons and their collaborators, much of the importance of

Mesoamerican nanacatism lay in the testimony it offered for the larger perspective emerging from their studies. Nanacatism uniquely and dramatically highlighted the significance of psychoactive fungi as cultural stimuli and their likely role in prehistory.

The simplest facts of the hallucinogenic mushrooms, of their use in native Mexico, or even their very existence, were largely unknown in the modern world until the public debut of the Wasson team's research in the May 13, 1957 issue of *LIFE* magazine. In an article titled "Seeking the Magic Mushroom," featuring sumptuous color photography and water color paintings, R. G. Wasson (1957) summarized his ethnomycological findings to date. He related how his studies in this subject had led him to the mountains of southern Mexico, where he and a few others become the first documented outsiders to be initiated into the rites of the sacred mushroom, by eating the fungi and experiencing their awesome effects along with their Indian hosts. For scientific study of the mushrooms, Wasson had enlisted the collaboration of the French mycologist Roger Heim, a distinguished expert in mushrooms, including species with pigmented spores, such as those used in the Mexican rituals.

(A technical note for the non-mycologist: one of the features of greatest significance for identifying mushrooms is the color of their spores, which can range from white through various hues, mostly earth tones, to black. These colors give an indication of species because they are constant, while other features, including those more obvious and readily observable, such as size, shape and coloration, can sometimes change with varying conditions in the habitat. This is the case for some of the sacred Mexican species, such as *Psilocybe caerulescens*, which "has a great variation in the form and color of the fruit body" [Guzmán 1983, 113].)

Based on specimens obtained in Mexico, Heim defined several new species, and in fact inaugurated an entire era of discoveries along related lines. By 1958, he and his colleagues had distinguished some thirteen species used by various Indian groups as sacred mushrooms, all but two of them previously unknown to science. New hallucinogenic species have continued to be described and named, and species concepts and relationships clarified, since Heim's groundbreaking work.

That the discoveries of the Wasson team caught the modern world unawares reflects a certain, almost charming, bygone naivety, a vacuum in scientific and cultural understanding relative to the sacred mushroom which was at a peak about a century ago. The circumstances by which the sacred fungi came to light have been related elsewhere. Excellent summaries include those of Stafford (1977), Pollock (1975), and Ott and Bigwood (1978), among others. Some of the most authoritative accounts however come from rather exclusive sources seldom seen, such as Wasson and Wasson (1957), Heim and Wasson (1958), and Heim et al. (1967). Of the latter works, the first is the only one written in English, and only 512 copies were printed. Similarly with Wasson et al. (1974); only 600 copies were produced. It may therefore be useful, especially with reference to the works just cited, to review these historic circumstances here, for they shed considerable light on the texts presented herein.

The Dark Ages, a Century Ago —the Safford Report

At the beginning of the twentieth century, nothing was known in any sciences or fields of study, anthropology, mycology, pharmacology, etc., of the sacred mushrooms of Mexico. Their very existence was virtually unsuspected. However, there were some little known traces of them from the early history of the New World. A few sixteenth century Spanish sources in Mexico had noted *teonanácatl* as mushrooms that caused an intoxication marked by visions. But in the first decades of the 1900's, there was no general awareness of, and no scientific answer to, these enigmatic early reports. They had never circulated widely, and only a handful of highly specialized experts had ever even heard of them. Knowledgeable leading opinion viewed these references, coming as they did from an era before the rise of modern science, with reservation. The whole matter was considered a red herring. True, as authorities of the time had verified, there was such a thing as peyote, also described for the first time in the early sources, which did indeed have effects much like those attributed to the sacred fungi. But peyote was a cactus, *Lophophora williamsii*, not a mushroom. Clearly, there was some confusion.

In terms of modern knowledge and understanding, these were the dark ages for the subject of the sacred mushroom. The few who had ever heard of any such thing mostly considered it fanciful nonsense. One leading authority on Mexican ethnobotany, William Safford, categorically asserted in an article (1915) that the old Spanish citations of *teonanácatl* were, in fact, merely mistaken references to peyote.

But Dr. Blas Pablo Reko, an Austrian physician who had lived in Mexico for some years, quietly disagreed with Safford's conclusion. He was also a student of Mexican plants and their native uses, and apparently had some inkling concerning the mushrooms. In a letter to Dr. J. N. Rose of the United States National Herbarium in 1923, he wrote that not only was Safford's conclusion wrong, but that the mushrooms are in fact still used in the Sierra Juarez of Oaxaca, adding that they grow on dung (Schultes 1940). Then in 1936 Reko's cousin, Victor A. Reko, wrote a book (titled *Magische Gifte*) in which he openly contradicted Safford's conclusion about *teonanácatl*. The book confused some facts and speculated in an indulgent manner, so it was not really a sound scientific work (Wasson and Wasson 1957, 238ftn.). But no one had ever disputed Safford about *teonanácatl* in print before. And in so doing V. A. Reko directed attention, for the first time in that generation, to the mystery of *teonanácatl*, almost forgotten since Safford's effort to dispel it more than two decades before, and now suddenly a subject of strong disagreement and lively controversy. In an unpublished note first reported on by the Wassons (1957, 238ftn), B. P. Reko wrote that in 1935 he had discovered the mushrooms in use among Chinantec Indian in Teocalcingo, as well as among Zapotec Indians in Santiago Yaveo, and that he had learned Chinantec names for them, *a-ñi* and *a-mo-kiá*. These early unpublished findings by B. P. Reko foreshadowed investigations to come.

1936, Huautla de Jiménez

Meanwhile during this same period, another Austrian native, an anthropologist of great importance in ethnomycology named Robert J. (Don Roberto) Weitlaner, was working among the Mazatec Indians of the town of Huautla de Jiménez, in the District of Teotitlán, State of Oaxaca. Huautla was to become a place of overwhelming significance for ethnomycology, a kind of ground zero where a process of discovery began, resulting in a chain reaction that made it world famous, and this is the point at which this fabled location enters the story. There, during Easter week of 1936, Weitlaner learned of the existence of Mazatec mushroom rituals from a close native associate and prominent local merchant. Moreover, he obtained a purported sample of the mushrooms, a first in terms of mycological collections. This was a major development, and constituted substantive evidence that Safford had indeed been wrong about *teonanácatl*. Realizing the importance of these specimens, Weitlaner sent them to B. P. Reko, thus vindicating the Reko refutation of "Safford's blunder" as Wasson (1980, 214) called it. In a single stroke, Weitlaner became the first to document the survival of mushroom cultism in Mesoamerica into modern times, and to secure the discovery with a voucher collection for study.

Early Mycological Studies by Richard Evans Schultes

In 1937, Reko sent specimens from Weitlaner's Huautla mushroom collection to the Botanical Museum at Harvard University, where they came to the attention of the young Richard Evans Schultes. Schultes was already on his way toward earning his legendary status as the world's leading ethnobotanist. The condition of the specimens Reko sent him was poor, precluding precise species identification. But to Schultes, the critical features that had been preserved clearly indicated it was a *Panaeolus* species. Members of this genus are black-spored mushrooms, generally known as inhabitants of livestock manure, especially horse and cattle.

Schultes thus became the first to scientifically study and identify a collection obtained as the sacred mushrooms of Mexico. And the results were interesting because *Panaeolus* species were implicated in a few cases of mushroom poisoning that had already come to the attention of both the medical and the mycological communities. These poisonings had been dubbed "cerebral mycetisms" in the literature, and they differed from other, more dangerous types of mushroom poisoning. The symptoms they exhibited were strangely consistent with those reported for the sacred mushrooms, so the notion of the Mexican specimens being *Panaeolus* species had a certain, distinct plausibility to it.

As if this were not enough, Reko also sent some specimens from Weitlaner's collection to Dr. Carl Gustaf Santesson in Stockholm for chemical analysis. Santesson's study indicated the presence of glycosides but not alkaloids, and he also investigated the physiological effects in frogs of an extract he prepared. Since the active ingredients of the sacred mushrooms are, as we now know, alkaloids (especially psilocybin and psilocin), this was obviously not the solution to the mystery. In light of discoveries made only later about the true status of *Panaeolus* in native ritual traditions, Santesson's negative findings are no surprise. Nonetheless, his report, which came out in 1939, was the first published attempt to crack the chemistry and pharmacology of the sacred mushroom.

1938-1940

Apparently intrigued with this incompletely solved riddle, and prompted by continuing reports of mushroom rituals following Weitlaner's discovery, Schultes undertook a trip to Mexico for ethnobotanical field studies in 1938, visiting various regions in Oaxaca including Huautla de Jiménez. With B. P. Reko, he obtained collections there representing at least two different species reportedly used in rituals, and carefully preserved them for study along with ethnomycological notes. Schultes' notes indicated that one of the collections, dated July 27, 1938, was regarded by the Mazatec as more important in their rituals (Heim and Wasson 1958, 184). With the collaboration of Dr. David Linder of the Farlow Herbarium, Schultes identified it as *Panaeolus campanulatus* var. *sphinctrinus* (Fr.) Bresadola. He now had a species level identification, and he presented his results in two definitive articles, the first in 1939 ("The identification of teonanacatl, a narcotic basidiomycete of the Aztecs") and the other in 1940 ("Teonanacatl, the narcotic mushroom of the Aztecs"). These articles by Schultes were the earliest published studies offering mycological findings on the sacred mushrooms from Mexico.

Beyond merely identifying mushroom specimens, Schultes (1940) presented a comprehensive report, weaving together various threads from the sixteenth century sources with the previous studies of Safford, Reko, Weitlaner and others. He integrated historic and anthropological data to give the clearest, most informative summary account of the subject yet. He reported Mazatec names for sacred mushrooms, attributing them to the *Panaeolus* species he had identified: "*t-ha-na-sa* (meaning unknown), *she-to* ('pasture mushroom') and *to-shka* ('intoxicating mushroom')" (1940, 435). Schultes had continued his 1938 field studies in Oaxaca in 1939, and determined that the Chinantec as well as the Mazatec used *Panaeolus campanulatus* var. *sphinctrinus* in rituals (Schultes 1940). He reported that in the western Chinantla (Chinantec country) this species was called *nañ-tau-ga*, and was used in various villages, including as a treatment for rheumatism in one. In the southeasternmost corner of the Chinantla he reported the use of mushrooms referred to as *a-ni* ("medicinal mushroom") and *a-mo-kya* ("medicine for divination"), but he did not obtain collections to establish their species identity. (The latter two names appear to be the same as those recorded by Reko in his private note, cited by Wasson and Wasson [1957, 238], referenced above.)

Meanwhile in Huautla, where Robert Weitlaner's observations had stirred so much interest with Reko and Schultes, anthropological research continued. There, on the night of July 16, 1938, just eleven days before Schultes obtained his historic collection of *Panaeolus campanulatus* var. *sphinctrinus*, Weitlaner's daughter Irmgard Weitlaner, along with her soon-to-be husband French ethnologist Jean Bassett Johnson, plus Louise Lacaud and Bernard Bevan, became the first documented outsiders in modern times to directly witness the use of sacred mushrooms by a *brujo* for curing by witchcraft, as Johnson called it (Wasson and Wasson 1957, 257). Contrary to some accounts, and based on authoritative sources, Robert Weitlaner was apparently not in attendance for this 1938 ceremony; nor had he witnessed any such ritual in 1936 when he gathered the first scientific "proof" of sacred mushroom ceremonies in Huautla. Johnson authored two articles in the next year describing the proceedings, of which the more substantial is "The elements of Mazatec witchcraft" (1939).

As an anthropological aside, Johnson's use of the term witchcraft here is technical. In popular usage, this same word appears in various contexts, carrying different but related meanings. In one usage it relates to an ambiguous identification, erroneous or otherwise, of indigenous European folk practices or teachings with concepts from Western religion, especially in the Christian tradition, of the devil and more specifically with his supposed human followers (real or imagined). In either case, the notion of witches is most often either taken seriously (with a positive or negative emphasis) or viewed as laughable, depending upon one's perspective. However, as used in the literature of anthropology, it carries different associations, and more detached value connotations. The term is applied to traditions in non-Western culture, connected with a vast ideational structure that includes components whose counterparts in modern society include such things as folklore, religion, mythology, medicine, psychology, entertainment arts, and fortune telling. A good starting point for the interested reader would be E. E. Evans-Pritchard's work on the Azande (1937).

The sundry native concepts referred to by anthropologists as witchcraft and witches vary substantially from one culture to another, and display interesting, even fascinating patterns. These are reflected in native vocabularies, where we discover a profusion of different terms with variously shaded meanings. A problem of interpretation inevitably arises as the English words and western concepts we attempt to use in understanding such native terms and realities collide head on with the very different ideational systems of these indigenous societies. Accordingly, one need not read any value judgments, positive or negative, into Johnson's use of the "w" word, as it is primarily a figure of anthropological speech. Similarly with Schultes (1940) who gives "witch doctor," as the translation for *brujo*, one of several Spanish terms widely used in Mesoamerica. One might compare the differing tone or emphasis of phrases in English, such as "witch doctor" versus "native healer," the latter expressing respect and the former with its skeptical or even derogatory undertow. Incidentally, the term "witch doctor" appears much more frequently in the ethnographic literature on African traditions than in that on cultures of the neotropics.

Johnson (1939) described the Mazatec mushroom traditions in some depth. He reported that in the ritual he observed, the *brujo* ate three mushrooms toward the beginning, and prayed to the Holy Trinity as well as native elemental powers, such as mountain-dwelling dwarves, the earth, stars, moon, sun, etc. Johnson also related Mazatec teachings about the fungi, for example, that it is the mushroom which speaks, not the *brujo*, when the latter is under its influence.

Like Schultes, Johnson mentioned the fact of sixteenth century reports of *teonanácatl*, but he also suggested that use of the mushroom did not exist among neighboring peoples such as the Zapotec. Further studies proved otherwise.

Although Johnson unfortunately obtained no voucher collections for scientific identification, he noted that a number of species were used by *brujos* throughout the Mazatec territory, including in San Cristobal Mazatlán, "the legendary capital of the Mazatecs" (1939, 133). These included a mushroom known in Spanish as *hongitos de San Ysidro*; another called *desbarrancadera*; and yet another, the smallest, called *tsamikindi* in Mazatec (Johnson 1939). Mycological studies since Johnson's time have revealed the two former species to be *Psilocybe cubensis*, and *Psilocybe caerulescens*, respectively. There is a third, smaller species frequently used among the Mazatec, *Psilocybe mexicana*, but Mazatec names for it as known from other reports do not seem to correlate obviously with *tsamikindi*. However, it may yet refer to *P. mexicana*, for the Wassons noted a diversity of names for the species used in Huautla. They considered "it was clear that the names were euphemistic escapes, metaphors rather than names, metaphors expressing respect and affection" (1957, 253). The Mazatec term usually reported for *P. mexicana* is *di-nizé*, as spelled by Ott (1976, 23). This means "bird," which almost certainly explains the origin of its Spanish name, also used by the Mazatec, *pajarito* or *pájaro*.

The basic foundation of ethnomycology was thus laid, from field studies to publications, by two major research efforts in the late 1930's, one purely anthropological and the other ethnobotanical. Both had focused on the Mazatec Indians in Huautla, following Weitlaner's findings there in 1936. By 1940, these investigations had borne fruit in the form of groundbreaking articles addressing the history, ethnology, mycology, and even chemistry of the sacred mushrooms. But these were only the first dim stirrings, for the great majority of new discoveries in ethnomycology still lay ahead.

One might have thought that the mystery of *teonanácatl* had been basically resolved by the findings of Schultes and Johnson, with only a few remaining details to be worked out. After all, their reports, the offspring of Weitlaner's work, together substantiated the sixteenth century accounts Safford had mistakenly dismissed. They established the reality of the sacred mushrooms and their traditional use in native Mexico beyond dispute. And they went further still by demonstrating that such practices had not gone extinct, but still lingered in some regions. Yet their findings, taken together, also indicated further questions.

Questions Remaining After 1940

In Huautla, Schultes had collected two different species regarded as sacred fungi, but identified only one, as reported in his publications. What was the other species, and how did it figure in native traditions compared with the one he had identified, *Panaeolus campanulatus* var. *sphinctrinus*? Johnson reported that at least three species were regarded as sacred in Huautla. He cited them by native names, but their mycological identities were unknown. These findings by Johnson harkened back to the sixteenth century report by Hernández, who also spoke of three different kinds of mushrooms the native Mexicans utilized for visionary effects. None of the indigenous names for sacred mushrooms reported from Huautla by Johnson (two of which were Spanish) matched any of those Schultes (1940) had cited. Clearly, there seemed to be more than one species of sacred mushroom, and little indication as to exactly how many or what they might be. To some, it appeared that findings to date might be only the tip of an iceberg. Schultes himself acknowledged this in his early reports, never claiming he had solved the riddle in its entirety:

> Although I have found no other mushroom used as teonanacatl in Oaxaca, nu-merous reports that there are several kinds of teonanacatl must be interpreted to mean that other species are actually used. It is probable ... that further ethno-botanical research will result in the discovery of other species which serve as divinatory-narcotics in southern Mexico (1940, 441).

Schultes probably could not have foreseen the extent to which he was cor-rect in this surmise, as was soon revealed by the investigations of others in the post-WWII era, especially the Wasson team. Even with the solid handle on the subject now in place, misunderstandings remained, and greater discoveries lay ahead.

Singer's Priority

The next major mycological discovery about the sacred mushrooms came with-out any fanfare. Although published in 1949, it received little attention because it was presented as but a tiny notice in passing, a mere speck embedded within an enormous, densely technical work where, like a needle in a haystack, it would take years to find.

In 1941, Schultes' second, unidentified collection of sacred mushrooms from Huautla, on deposit in the Farlow Herbarium at Harvard University, was studied by agaricologist Rolf Singer. He determined it to be *Stropharia cubensis* Earle, a species that had been described in 1906 based on specimens collected in Cuba (hence the name *cubensis*, which in Latin means "from Cuba"). This was a significant new conclusion.

In the course of his microscopy with this collection, Singer noted some technical features of this species that tended to contradict its original placement in *Stropharia*. On that basis, he re-classified it, transferring it to the genus *Psilo-cybe*. He thus coined the new name *Psilocybe cubensis*, and amended the herbar-ium labeling of the specimen accordingly. With these findings, Singer became the first to conclude that a *Psilocybe* species, and specifically *P. cubensis*, was used as a sacred mushroom by the Mazatec. He first reported this result eight years later, in the first edition of his massive book, *The Agaricales in Modern Taxonomy* (1949). However, the mention Singer made therein of this finding is so brief and incidental, one is left with the impression that he may not have fully realized its significance at the time.

The quote of interest—there are actually two, the other (Singer 1949, 506) being even briefer, and forgoing any mention of species—occurs in a section on the genus *Panaeolus*. It reads:

> *Panaeolus sphinctrinus* and *P. papilionaceus* are used as intoxicating drugs in Central America by certain Indian populations, together with *Psilocybe cubensis*. In large doses (i.e., 50-60 specimens) they are poisonous. *Panaeolus* occasionally appears as a weed fungus in mushroom beds, but the damage inflicted is probably negligible (1949, 472).

As Wasson later related (1963, 56), he did not learn of Singer's identification of *Psilocybe cubensis* as a hallucinogenic fungus in Mesoamerica until 1957, when Singer apprised him of it in person on the one and only occasion when the two met (for a photograph of this momentous event well worth a thousand words see Guzmán 1990, 86). Wasson had already spent several years researching firsthand in the field to learn what species of mushrooms were used in the Indian rituals, along with an international team of experts he had by this time assembled. No doubt he would have been urgently interested to know of Singer's finding sooner, as one clearly senses from some of his published remarks on this twist in the history of ethnomycology (see Wasson 1963, 56).

Genus and species

Psilocybe is a genus of mushrooms in which the spores in mass display a purplish tint, darker or lighter, and the cap is usually slippery or sticky when moist (a feature not readily discernible under dry conditions). *Psilocybe cubensis*, or *Stropharia cubensis* as the Wasson team continued to call it, was the species referred to in Huautla by the Spanish name *hongitos de San Ysidro*, as reported by Johnson in 1939. Heim and Wasson (1958, 53) give this name as *honguillo de San Isidro Labrador*. Among the *Psilocybe* species used as sacred mushrooms in Mexico, this one is distinguished by its occurrence in cattle and horse manure, a habitat strangely enough less reminiscent of other sacred species in the genus, and more like that of *Panaeolus campanulatus* var. *sphinctrinus*. The other ritually utilized *Psilocybe* species occur in other habitats, such as meadows or marshes or rotting sugar cane refuse or cultivated fields, or disturbed grounds such as landslides, and so on. Certainly, there are other species in the genus that occur in manure, for example *Psilocybe coprophila*, a very common mushroom. But it is not used in rituals nor does it have any significant psilocybin content.

Psilocybe cubensis made itself known early on in the course of research on the sacred mushrooms, indeed it was the first in its genus to do so. Since then, it has continued to acquire a fascinating history in modern times, winning many distinctions, particularly within the counterculture. For example, this is the species which has made pasturelands of the Deep South legendary for "magic mushroom" hunting (although other hallucinogenic taxa, for example *Copelandia* species, also occur in some of these). Likewise, it is this species that, by its vigorous viability on growth media such as sterile agar and cereal grain, has made clandestine cultivation of psilocybin mushrooms in the modern world a widespread reality, beginning in the 1970's. In connection with this circumstance, it has also become the species most often revealed by examinations of hallucinogenic fungi obtained by illicit means, among those that prove to be genuine psilocybin mushrooms and not *faux* products, as were more often encountered before the rise of home *Psilocybe cubensis* cultivation (Anonymous 1973; Pollock 1975, 76). Considering its habitat in cattle manure, and given its status as one of the sacred Mexican mushrooms, perhaps this species somehow exudes a kind of unique, uncanny fascination. One may be almost reminded of the lotus which emerges from the most stagnant water yet is unsullied by it, delicate and beautiful, a symbol in Eastern tradition of transformation and purity.

As a mycological aside, different authorities over the years have subscribed to changing, differing definitions of various genera of mushrooms. Depending upon which definition one follows, species may be classified in one genus, or some other. Mushroom taxonomy is like international diplomacy, a matter of formal recognition, and various species are thus occasionally subject to conflicting expert opinions about how they should be classified. *Psilocybe* and *Stropharia* are classically regarded as closely related genera, recognized in the same family, the *Strophariaceae*, and the name *Stropharia cubensis* is still considered valid for this species. Meanwhile, as noted by Stamets (1996), the mycologist Nordeloos has more or less annulled the genus *Stropharia*, transferring the entire catalog of species traditionally classified there into a resultantly expanded genus *Psilocybe*.

Enter the Wassons

The debut of R. Gordon and Valentina P. Wasson into the subject of the sacred mushrooms of Mesoamerica came in the early 1950's. As a couple with contrasting cultural backgrounds, they had become aware early on, since their marriage in 1927, that they had sharply differing personal dispositions toward mushrooms and fungi in general. They also noticed these contrasting dispositions seemed to correspond to basic attitudes of their respective peoples, she being of Russian descent and he an American, born in Montana. In English tradition, wild fungi on the whole are customarily regarded as "toadstools," objects worthy of being shunned. But in Eastern Europe, wild mushrooms are admired, accorded affection and charming folk stories, and avidly gathered from the wild for the kitchen. The Wassons took keen notice of this opposition, and it became the inspiration for their wide-ranging studies on fungi in human affairs, a research endeavor for which they coined the word ethnomycology. Their goal was to discover the basis, the explanation, for this strange schism of mycophobia and mycophilia in the tree of cultural history and evolution. (See in this volume Benítez, "The Hallucinogenic Mushrooms," for his account of the Wassons' contributions to our knowledge about this subject.)

The Wassons' inquiries into this riddle of the ambivalent connection between fungi and humanity had begun as far back as the late 1920's. By the early 1950's, they had traced the dim outline of this rough beast, in the process developing an extensive network of correspondents and contacts over the many years of this research. Among the matters they learned about was the use of fly agaric (*Amanita muscaria*), a psychoactive mushroom of properties and chemistry very different from that of the sacred Mexican fungi. Traditionally feared as deadly throughout much of Europe, early observers in Siberia noted to their amazement that various reindeer-herding peoples who inhabit some of its remote regions ingest the fly agaric as a hallucinogen in shamanic rituals.

Based on everything they were able to discover, the Wassons speculated that the cultural split between mycophobia and mycophilia was an artifact or echo of a time in prehistory when fungi must have been worshipped or more widely involved in religious life. The traditional Siberian use of fly agaric was, in this view, a sort of ethnographic fossil, a surviving relic of the ancient past. This idea differs sharply from some analyses of these Siberian traditions, most notably that of Eliade (1964). But it seemed to fit well with the general adoration of fungi in mycophilic cultures, and also with the mycophobic aversion to them, understood as a remnant expression of the *tabu* nature of the sacred, "the aftermath of the emotional hold of those mushrooms on our own ancestors" (Wasson and Wasson 1957, 375). For a definitive and persuasive rebuttal of Eliade's interpretation of the fly agaric in Siberia, see Wasson (1968, 326-334).

Early on, the investigations of the Wassons had focused primarily on the Old World. But in September, 1952, two items arrived in the Wassons' mail, by "almost the same post" (1957, 275) which redirected their attention. First was a letter from a correspondent in Verona describing a stone artifact from Mesoamerica shaped like a mushroom, which he had seen in the Reitberg museum in Zurich. Then their friend the celebrated author Robert Graves sent them an article he had found summarizing the state of knowledge at the time about ritual use of hallucinogenic fungi in Mexico and elsewhere, including the findings of Schultes and Weitlaner (Heizer 1944). This was the first the Wassons had ever heard of *teonanácatl*. These two items together constituted their first inkling of the Mesoamerican mushroom connection, and it sparked their interest mightily.

The First Wasson Expeditions to Mexico (1953-1955)

The Wassons quickly realized that the Mexican sacred mushrooms would likely be of overwhelming relevance for their theory about mycophobia and mycophilia. They also understood there were questions already raised by others which had not been answered, such as what species other than *Panaeolus campanulatus* var. *sphinctrinus* were involved. They wrote to Weitlaner, and to Reko, and were soon directed for further information to a missionary living among the Mazatec Indians, Eunice V. Pike. After correspondence with Pike, in which she revealed tantalizing details about persisting mushroom rituals, the Wassons undertook their first expedition to Mexico and Huautla de Jiménez, in 1953. This was the first of many excursions by R. G. Wasson and associates to Mesoamerica to investigate the sacred mushroom complex.

With assistance from their new, experienced senior colleague Robert Weitlaner, the man credited with the rediscovery of the Mexican mushroom rites in modern times, the Wassons succeeded in 1953 in witnessing a mushroom ritual, conducted by Aurelio Carreras, a *curandero* (a general term in Mexico for a native diagnostician and healer). The following year, at Robert Weitlaner's invitation, R. G. Wasson returned to Mexico with his photographer and friend Allan Richardson. Walter Miller, an expert in the Mixe language and culture, joined in to assist them in their studies for this round (see Miller, "The Mixe *tonalamatl* and the sacred mushrooms").

Then in 1955 the Wasson team mounted a third expedition which led to an even more momentous breakthrough, referred to by Wasson as the mushroom agape. This was the first documented instance of non-Indians, in this case Wasson and his associates, being invited by the Indians to participate in the ritual by ingesting the fungi and undergoing their effects along with them. The ceremony was conducted by the Mazatec *curandera* María Sabina, who ultimately achieved a measure of international renown as a result of these circumstances, but also endured some bitterly harsh consequences (Estrada 1981). The Wassons had been already preparing to publish the findings from their studies before this journey, but after this development they were bursting at the seams. In particular, the personal impact and power of the effects of the mushrooms came to them as a startling revelation.

1957: The Public Unveiling

In 1957 the Wassons summarized their preliminary results from years of exhaustive multidisciplinary studies, in both an abridged, popular form in *LIFE* magazine (Wasson 1957), and in their larger, more comprehensive work, the seminal two volume set *Mushrooms, Russia and History* (Wasson and Wasson 1957). The latter is an exclusive, rarely seen, exquisitely wrought tome, printed in Verona, Italy, of which only 512 copies were produced. Volume II presents the Wassons' early Mesoamerican studies in detail, covering findings from their expeditions undertaken in 1953, 1954, and 1955. In contrast to the limited availability of this book, *LIFE* had an enormous circulation and reached a wide general audience who read Wasson's account of the sacred mushrooms and viewed the impressive illustrations with understandable intrigue and amazement. This was an unprecedented kind of feature, nor has its like been seen since.

Wasson admired a certain richness and breadth of study he had noted in the French mycological journal, *Revue de Mycologie*. While in Paris in 1949, he made the acquaintance of its editor, the French mycologist Dr. Roger Heim. Heim thus became Wasson's mycological mentor and expert advisor on technical and scientific aspects of fungi for Wasson's ethnomycological pursuits. When the matter of the Mexican sacred mushrooms arose a few years later, Wasson naturally solicited Heim as his collaborator to study and identify collections of the species used in rituals (Wasson et al. 1986, 20). By 1957, this branch of the investigations had borne fruit. Heim had identified *Stropharia cubensis*, as well as another species, likewise already known to mycologists from prior description. It was *Psilocybe caerulescens* Murrill, originally discovered from Alabama according to the records. Beyond these old and vaguely familiar fungal faces, Heim had also described and named several new species, such as *Psilo-cybe mexicana* Heim, and *Psilocybe aztecorum* Heim. All four of these species were illustrated in the 1957 *LIFE* magazine article in elegant, stylish water color renditions by Heim. But not all the new species the team discovered had been published yet, as of the debut of *Mushrooms, Russia and History* that same year.

This information gap was closed in large measure the next year by Roger Heim and R. Gordon Wasson et alia in another praiseworthy tome, *Les Champignons hallucinogènes du Mexique* (1958), published in Paris. Here, a full and unusually lavish mycological presentation detailed thirteen species Heim had determined as sacred mushrooms in Mexico. Eleven of these were in the genus *Psilocybe*; with a twelfth, *cubensis*, reckoned by Singer also as a *Psilocybe* (although Wasson and Heim retained it in *Stropharia*). The thirteenth, *Conocybe siligineoides* Heim, was a brown-spored species. Like the latter, most of the sacred *Psilocybe* species were previously unknown to mycologists, described and newly named by Heim. As reflected in the pages of *Les Champignons hallucinogènes du Mexique*, understanding not only of native Mexican culture, but also of the genus *Psilocybe* had been dramatically advanced by the Wasson team's studies of nanacatism.

Meanwhile on the other side of the Atlantic, also in 1958, a somewhat parallel work appeared in the pages of *Mycologia*, the journal of the Mycological Society of America. Rolf Singer and his colleague Alexander Smith published a taxonomic description and classification of all species in the genus *Psilocybe* known to be hallucinogenic, along with others suspected to be so because of a telltale characteristic: a tendency of the fresh mushroom to bruise cerulean blue at the site of any rough handling or injury, on either the cap or stalk. All the sacred species described thus far shared this blue-bruising tendency. And from other parts of the world, there were others in the genus that shared this feature, as known purely from mycological studies, minus any anthropological aspects. Singer and Smith (1958b) predicted all these species would have hallucinogenic properties, and designated a sub-generic group for them, the section *Caerulescentes*. The sacred species from Mexico were classified there, including a few that occurred not only in Mexico but also in the United States. For example, Singer let one cat out of its bag by noting he had seen *Psilocybe cubensis* in Florida (Singer 1958: 250). This was a resounding early note that was sooner or later heard, as subsequent history suggests, by an increasingly wide audience.

But a bone of contention also emerged from this article when it developed that Singer and Smith (1958a) claimed as theirs to name, by prior publication, a new species which the Wasson team had been the first to collect and recognize as new. Heim was in the process of having this species designated *Psilocybe wassonii*, in honor of Wasson, and indeed that is how it appears in *Les Champignons hallucinogènes du Mexique* (1958). But in the pages of *Mycologia*, Singer and Smith beat Heim to the finish line, by almost a month. Procedures for designating new species are subject to certain conventions, arrived at by agreement within the international scientific community, not all of which Heim had yet satisfied (Guzmán 1983, 262). Under the formal rules of nomenclature, Singer and Smith's name, *Psilocybe muliercula*, thus took precedence. This was another seed of discontent sown between mycologists in the United States, and the team Wasson had recruited (Ott 1976, 31n.). Under such circumstances, and following proper taxonomic form, citations of this species often include both of these published names, as follows: *Psilocybe muliercula* Singer and Smith (= *Psilocybe wassonii* Heim).

Heim and his associates documented a diversity of species used as hallucinogens in Mexico, as did Singer and Smith. But both groups noted a conspicuous absence: the genus *Panaeolus*, especially including the species Schultes had collected and identified from Huautla, did not figure among the mushrooms used in native rituals. Schultes' mycological findings, the results of his work with microscope and technical references, have never been in dispute. Singer and Heim studied, independently, the 1938 collection he had determined as *Panaeolus campanulatus* var. *sphinctrinus* and both confirmed its identity as such. What neither Singer nor the Wasson team could confirm was the native status of this species as a sacred mushroom. Indeed, taken as a whole, investigations in the modern era have revealed no indigenous use of *Panaeolus* in Mexico as a sacred mushroom.

In fact, *Panaeolus campanulatus* var. *sphinctrinus*, or *Panaeolus sphinctrinus* as Singer (1949) called it, is not a hallucinogenic species. As reflected by scattered medical reports of mushroom poisonings linked to *Panaeolus*, there are indeed a few species in this genus that contain psilocybin, but this is not one of them. There has been a slightly treacherous history of misidentifications, even by expert mycologists sometimes, between the genera *Panaeolus* and *Psilocybe*, the latter being misidentified as the former, as noted by Gastón Guzmán (1990, 87). Schultes' misinformation may have resulted from a native informant who, logically, could have made a similar mistake, especially considering the dung habitat of *Psilocybe cubensis*. Field guides for mushroom hunters often warn readers of the difficulties of identification for some of these "little brown mushrooms" or LBMs as they have sometimes been designated. Their critical features tend to be obscure, in some cases requiring extensive technical knowledge and microscopy to conclusively demonstrate. A telling clue to this seeming mystery may lie in an observation by the Wassons pertaining to the species used in Huautla, that they "included many small mushrooms, some with black and others with purple-brown spores. The Indians esteem them highly ... They belong to various species but the Indians do not distinguish among them mycologically" (1957, 254). The mycologically astute reader will note that the difference in spore color indicated here is consistent with the distinction between *Panaeolus* and *Psilocybe* species. The Wassons add that these mushrooms grow in pastures, typical for *Panaeolus*, and that this mixture of species was referred to in Spanish as *angelitos*—a specific native Mexican colloquialism for *Psilocybe mexicana*.

The discoveries of the Wassons and their colleagues, as they became public starting in 1957, caused something of a sensation. They soon began to take on a life of their own, becoming incorporated into a larger course of history that was only just beginning to emerge in the second half of the twentieth century. The roots of this emergence lay in various places, including Basel, Switzerland where, in 1943, the effects of LSD were quietly discovered by Albert Hofmann.

Hofmann had developed this semi-synthetic compound in 1938 from alkaloids extracted from ergot (*Claviceps purpurea*), another fungus, not related to the Mexican sacred mushrooms, with its own long, intriguing history and reputation in the Old World. This history had brought it to the attention of Sandoz, the pharmaceutical firm where Hofmann was employed, as a potential source of new active compounds. But no one could have anticipated that research on ergot alkaloids would lead to the discovery of a substance with effects such as those of LSD, especially as manifest in such tiny amounts (Hofmann 1980).

The discovery of psilocybin and psilocin

Beyond merely documenting, describing and naming the species used, with herbarium collections for scientific legitimacy, Heim and his mycological colleague Roger Cailleux isolated living cultures of species they studied, and grew specimens on sterilized nutrient media in their laboratory in Paris. One of their goals for this phase of research was to supply fresh sacred fungal produce in quantities sufficient for chemical studies to determine what compounds in them were responsible for their hallucinogenic effects.

Heim and his associates were cognizant of the recent discovery of LSD, and aware of the analogy between its effects and those of the sacred mushrooms. On this basis, Heim wrote to Hofmann in 1957, soliciting his collaboration to unravel the chemistry of these species. Hofmann accepted the invitation, and Heim thus began sending him dried fungal material cultivated in Paris.

For Hofmann, the first step was simply to confirm whether the cultivated specimens possessed the same hallucinogenic activity as those from nature, and would therefore not be a waste of time for chemical investigations. Unfortunately, studies using living animals proved useless for this purpose. Hofmann was finally obliged to ingest some of the material himself, in effect using his own central nervous system as the instrument for determining its potency. After the outcome proved positive, Hofmann proceeded to extract several crystalline compounds from the fungi, separating them by paper chromatography. Through experimentation with samples of the purified, separated compounds, he soon established that the hallucinogenic effects of the mushrooms resided in two alkaloids, which he named psilocybin and psilocin. This simple method had its precedent in the historic determination of mescaline as the hallucinogenic constituent of peyote by Heffter in 1896 (Anderson 1980, 119).

In the next phase of his work, Hofmann determined the chemical structures of these new compounds, which proved to be tryptamine derivatives. To accomplish this task, larger quantities of the purified compounds were needed. The raw fungal material he used in this step was a supply Heim sent, not of mushrooms, but of sclerotia of *Psilocybe mexicana*. The sclerotia were an entirely different type of multicellular fungal structure, unique and unfamiliar from previous experience. They formed abundantly in cultures, but only of this one species out of the several which Heim and his co-workers had grown.

Unlike the mushrooms themselves, which are delicate and ephemeral, and have a distinct pileus and stipe, the sclerotia (sing: sclerotium) are dense, compact, persistent structures that lack the familiar cap-and-stalk body plan of a mushroom, and produce no spores. They were first described and illustrated in 1958 in *Les Champignons hallucinogènes du Mexique*. As was later discovered, sclerotia are also produced by a very few other hallucinogenic species of *Psilocybe* (Guzmán, 1978). Sclerotia of *Psilocybe* contain the same active ingredients as the mushrooms, and are known mainly from cultivation studies. However there is one isolated report (Ott and Bigwood 1978, 94) that the Mazatec are acquainted with sclerotia from the natural habitat, of *Psilocybe caerulescens*, a species which oddly enough has never been reported to produce these structures in culture. The Mazatec reportedly refer to these sclerotia as *camotillos* ("little sweet potatoes") or *raíces de los derrumbes* ("roots of landslide mushrooms," i.e., of *Psilocybe caerulescens*). This finding is backed up by a rare voucher collection made by Wasson colleague Jonathon Ott, from the Huautla de Jiménez region in 1975, securely identified by leading authority Gastón Guzmán (1978, 640). For photographs of cultivated sclerotia of *Psilocybe mexicana*, and *Psilocybe tampanensis*, a rare, closely related species from Florida, see Stamets and Chilton (1983).

A technical aside: mycologists use the term sclerotium widely in reference to any of various different kinds of hard, compact, durable structures produced by diverse fungi, especially if they do not form spores. These are by no means necessarily equivalent to the sclerotia produced by *Psilocybe mexicana* and its close relatives. Ergot, *Claviceps purpurea*, provides a convenient example. This species typically transforms some of the kernels of the grains it parasitizes into "spurs," dark, enlarged structures representing a dormant winter stage in its life cycle. These spurs are technically designated as sclerotia. In the spring, with warmer temperatures and rain, fleshier, more ephemeral structures called *stromata* (singular: *stroma*) sprout from them. The stromata are the spore-producing stage of *Claviceps purpurea*. Despite the identical terminology, sclerotia of *Psilocybe* species, of ergot, and of various other fungi are not really equivalent or comparable, other than all being hard, dense bodies.

Doors of Perception

Hofmann and his collaborators in Basel and Paris soon determined the organic structures of psilocybin and psilocin, and published their findings in 1958 and 1959. With the chemistry of the sacred mushrooms now clarified, understanding of nanacatism in Mesoamerica was reaching a new and more comprehensive level. A few decades earlier, the only compound known to science with comparable effects was mescaline. But with the discovery of LSD, plus the new alkaloids from *Psilocybe* species, pharmacologists were beginning to realize the existence of a whole class of compounds with a strange activity resembling that of mescaline.

As revealed by further investigations, the effects of these substances were manifested primarily within the stream of consciousness, targeting the complex, higher processes of cognition and affect, and inducing profound, even baffling alterations of perceptual function and response. These effects could be described or interpreted by analogy with either schizophrenic symptoms, or with states of creative and spiritual inspiration, depending upon their specifics and one's point of view. Furthermore, experiences with these drugs seemed to have a lingering effect afterward on many subjects, such that their entire worldview seemed to undergo a deep and persistent shift, especially from philosophical materialism or rationalism to a more contemplative, even mystical perspective. There was thus both an acute dimension to the effects of these substances, manifest at the time of intoxication, as well as the aspect of a long term outcome to account for. This was remarkable, and unprecedented in pharmacology.

A compelling personal account of these complex effects was that of the well known author and intellectual Aldous Huxley, whose work became another root of the larger history into which the discoveries from Mexico were gradually assimilated. In 1954, with the publication of his classic book *The Doors of Perception*, Huxley detailed remarkable experiences of consciousness alteration and mystical transcendence which resulted when he took mescaline, provided to him by Dr. Humphry Osmond, a psychiatrist investigating the drug. Huxley and Osmond agreed that the effects of mescaline were provocative and challenged comprehension, baffling the mundane, routine categories of thought and understanding habitually taken for granted on a daily basis.

Indeed, the efforts of most subjects to describe these effects often resulted mainly in excited babbling and incoherence, reinforcing the similarity to psychotic states perceived by some investigators. For these investigators, the most compelling hypothesis so far adduced held that substances such as LSD and mescaline were psychotomimetic, meaning they simulated what goes on in the mind, and presumably the brain as well, in connection with psychosis. However, Huxley soared like an eagle above the trap of verbal confusion, rising to meet this seeming expository challenge of the intangible, intellect-boggling phenomenology of mescaline intoxication that underlay it. The incoherence of some subjects was nowhere to be found in his report. He made compelling sense by explaining his experiences, and the impression of revelation they imparted, by using insights from religious teachings and literature of the Far East. As Huxley clearly demonstrated, there was far more to such experiences than a breakdown of mental function, notwithstanding citations in the research literature to model psychoses, and the frequent similarity in appearance between true inspiration and abject insanity, as often acknowledged by conventional wisdom. His book was soon followed by *The Joyous Cosmology*, an equally interesting and impressive offering along related lines by another noted scholar of Eastern religions, Alan Watts (1962).

Traditions such as Buddhism, Hinduism and Taoism were richly furnished with teachings, not all easily grasped, relating consciousness to spirituality. Most of these ideas seem strange by the standards of Western philosophy and religion, but for many readers they have proven unusually well suited for understanding and interpreting the subjective effects of LSD-like substances, not in pathological terms as a model psychosis, but as a kind of radical expansion of awareness. One of the intriguing aspects of these effects is that they seem to operate to a great extent directly upon thought and consciousness, faculties developed to a uniquely high degree in the human species. Huxley was one of the few people sufficiently astute, and educated enough in these non-Western traditions, to realize the relevance of Eastern religion for interpreting the effects of mescaline, and he was the first to drive this point home with eloquent, cogent discussion.

The Doors of Perception remains a monumental classic, and it is difficult to overestimate its historic importance. Huxley powerfully defended the experiential validity and authenticity of the noetic phenomena induced by mescaline, such as mystical experiences. He spoke for realms of transcendence he found lying unsuspected behind glib and misleading designations such as hallucinogen or psychotomimetic. In so doing, Huxley fired the first shot in what became the battle over LSD and other hallucinogens in the 1960's. (For an excellent, vintage discussion that takes a hard-boiled look at this controversy and its connection with Eastern religions, see *The Private Sea* by Braden, [1967]; for a superb retrospective account and analysis of LSD and the controversy surrounding it in the 1960's see *Storming Heaven* by Stevens [1987].)

The interest stirred by *The Doors of Perception* and early research on LSD gave a special added impetus to the investigations into the sacred mushrooms of the 1950's. Various investigators of LSD and similar agents were realizing the challenge for theoretical description posed by their effects, and a discussion was joined about how they should best be designated. A suggestion made by Huxley which never gained any great currency was phanerothyme ("revealing the soul or spirit"). In correspondence with him in 1956, Osmond offered a counter-suggestion: psychedelic, meaning "mind-manifesting." This was the origin of the word, admirably unassuming as to interpretation, which became an icon of the era that soon followed (Stafford 1977, 5-6).

Thus, between Huxley's landmark book in 1954, and the discovery of LSD's effects in the previous decade, the investigations into the sacred mushrooms of Mexico found a ready framework into which the discoveries of the Wasson team, now including Albert Hofmann, fit. Some sense of the first inkling of a theoretical connection between LSD and the sacred fungi is evident in the following quote by the Wassons, from 1957. It reflects the long-since-lost innocence of that era, which, by twenty-twenty hindsight, appears so remarkable in light of what was to follow in the 1960's:

> From the fungus known as ergot Swiss pharmacologists have recently isolated an alkaloid that causes massive psychic reactions in human beings, including hallucinations that duplicate with astonishing fidelity the testimony of our old Spanish writers. Experiments with this alkaloid in England and America seem to open up promising vistas for its use in the treatment of psycho-neurotic disorders (Wasson and Wasson 1957, 241).

The elucidation of the sacred mushrooms was momentous in itself, but it was becoming gradually incorporated into seismic historic rumblings even deeper. Sociocultural developments of vast proportions were beginning to stir and upwell with tidal force.

Slowly, the mushrooms became enmeshed within the fabric of what developed into the psychedelic counterculture of the 1960's, and beyond. Unheard of by the general public before 1957, the magic mushrooms were on their way to becoming a significant fixture in popular knowledge and awareness. A psychedelic dawn was breaking on the horizon of modern history, and the mushrooms would play an important role. An emblematic reflection of their hold came in 1974 with the advent of *High Times* magazine, a sort of popular journal of illicit recreational neuropharmacology that achieved a considerable commercial circulation. The cover of the first issue depicted a young woman, of presumably post-hippie persuasion, eating a mushroom. From what we can see in the photograph, the specimen shown may have simply been an edible *Agaricus* obtained from a local grocery. But its intended representation as a magic mushroom was not lost upon the target audience. (For an entertaining and informative history of *High Times*, see *High Times Greatest Hits: Twenty Years of Smoke in Your Face,* published in 1994 by the Trans-High Corporation, the editors of *High Times*).

A key circumstance relating to the place of the sacred mushrooms in the contemporary milieu concerns the historic role of Timothy Leary in the rise of the psychedelic 1960's. His name became famous, or notorious depending on one's perspective, for his public advocacy of LSD. But his interest in such substances actually originated not with LSD, but rather with his first psychedelic experience, which involved sacred mushrooms in Mexico in 1960. The mushrooms played no direct role in the discovery of LSD by Hofmann. But the psychedelic 1960's and the fascination with LSD clearly owed a great deal on the one hand to Huxley's *Doors of Perceptions*, and on the other to the romance of the magic mushroom, especially as publicized in 1957 in the pages of *LIFE* magazine. The publicity surrounding the mushrooms got a second boost in 1961 through the magic of television, courtesy of ABC-TV and the creators of *One Step Beyond: The Sacred Mushroom* (see Chapter Seven).

During the 1960's, amid the psychedelic social upheaval of that era and the sensational journalistic coverage it received, ethnomycological research proceeded, for the most part minus the sound and fury of the media horn, which focused more intensively on LSD. In 1967, significant new findings from Mexico were presented in another landmark book published in Paris, *Nouvelles Investigations sur les champignons hallucinogènes*, by Heim and Wasson, et al. Soon after, Wasson brought out his classic book about the fly agaric, *Soma: Divine Mushroom of Immortality* (1968), in which an intriguing link was proposed between practices of Siberian reindeer-herding peoples and the riddle of *Soma* from the RgVeda of ancient India. Wasson's hypothesis, that *Soma*, the mysterious sacred intoxicant of the Vedic religion was in fact fly agaric, has been supported since it was first proposed by interesting evidence more recently discovered. For example, later studies demonstrated that when the use of *Soma* (whatever its species identity) disappeared many centuries ago in India, a mushroom of the genus *Scleroderma* was initially used as a ritual substitute for it (not for purposes of ingestion, but in a ceremony requiring that a ceramic vessel containing it be fired in a kiln) (Wasson et al. 1986).

By the 1970's, Osmond's word psychedelic had been severely bandied about, and applied in all kinds of popular, extraneous contexts such as music, art and literature. This in itself reflects the prominence which LSD, and to a lesser extent psilocybin, had achieved. But the swirl of events into which the mushrooms were drawn eventually reached such proportions that Wasson and some of his collaborators, perhaps wishing to distance themselves, felt compelled to abandon the use of the word psychedelic. They proposed instead a new term, entheogenic, to describe the mushrooms and their effects. This word approximately means "producing the god within" and aptly acknowledges the indigenous concept of the mushrooms as sacred (Wasson et al. 1986, 30-31). Its meaning also corresponds well with various types of transpersonal experiences sometimes induced by such agents, as described by Huxley and many others since, including Albert Hofmann and the Wassons. Ethnomycological investigations in the 1950's, proceeding from the foundation laid in the 1930's by B. P. Reko, Robert Weitlaner, R. E. Schultes and J. B. Johnson, thus came to impact deeply upon modern history, helping to shape its course in ways no one could have predicted, indeed, becoming part of that shape. History turned on these discoveries, even as these discoveries turned on history.

However, as revealed through ethnomycological investigations, the sacred mushrooms of Mexico remain a fascinating subject matter unto themselves, even without taking into account their greater historic impact in the modern world. Simply as expressions of native tradition and belief in Mesoamerica, nanacatism, with its basis firmly in fungi, stands as a uniquely rich and interesting phenomenon. Our modern perspective on it, coming from an external cultural viewpoint, has undergone some refinements since the first Western contacts with this tradition centuries ago. Contemporary understanding of the effects of active substances such as those of the mushrooms and peyote can now provide for a different analysis, one that was simply not possible for the Spaniards in an era before the rise of science.

A number of studies on the effects of LSD and related compounds have lent support to the views of Huxley and Watts, indicating that a "transcendent state" can result in some subjects, closely resembling spontaneous religious experiences, especially of a mystical nature, at least as a matter of phenomenological description (Masters and Houston 1966, 257). A particularly striking example of such research, specifically using psilocybin, was the noted "Good Friday" study by Walter Pahnke in 1962. The use of the term mystical here is technical and fairly specific, involving certain definite experiential criteria, versus a somewhat looser application of the term one finds in literature other than that of theology and comparative religion, such as anthropology (Pahnke and Richards, 1969).

It seems a safe prediction that the sacred mushrooms will continue to command the interest of a wide audience, and perhaps the apprehension of various authorities in the society, religious and secular. But meanwhile, findings that have verified the mysticomimetic effects of compounds such as those in the mushrooms can help yield a more insightful and humane understanding of phenomena such as nanacatism, one that is considerably advanced over antiquated notions of drugged ritual frenzies. Such findings yield a view of the peoples who espouse such traditions not as backward savages indulging in fundamentally "superstitious" activities, but rather as dignified human beings manifesting a deeply religious and spiritual nature. Furthermore, their practices as such actually make considerable sense, in light of what we now know about the effects of the mushrooms. In the Mesoamerican cultural setting the mushrooms have a practical, recognized value, and the deeply rooted ritual context offers an important anchor against the ambiguities of their considerable power.

Similar considerations may shed light upon the recent history of modern society and the role of psilocybin mushrooms and LSD therein, especially with regard to its countercultural aspects. It seems a common assumption that illicit use of such agents reflects an inherent negation or loss of authentic traditional values, a sort of post-juvenile delinquency expressing a lack of direction, or generation gap. But taking the mysticomimetic aspects of mushroom intoxication into account, another interpretation, with less culturally pathological implications, is possible. Tentatively, we may propose that the prospect of direct access to profound, even transcendent dimensions of human experience, vital in some way at a psychospiritual level spanning the person and the cosmos, and closed off from normal, routine experience day to day, is probably what generates and sustains interest in these substances within the counterculture. The interest we see there may primarily reflect that the perennial search for meaning is alive and well, which is perhaps less cause for lament than celebration, even relief, all things considered.

Such an explanation obviously requires a somewhat flexible or open-ended, even pluralistic theoretical framework. But what else could better explain the apparent willingness of many people to personally experiment with ingestion of such agents, considering the legal risks posed, as well as the hazard of traumatic stress residing within psychedelic experience? From this perspective, there is perhaps a basis for a sympathetic understanding of the contemporary interest and even enthusiasm for the effects of the sacred mushrooms—however lacking in the guidance and wisdom of well-informed tradition—which might otherwise seem understandable only as folly or misguided self-indulgence.

In the Mesoamerican context, nanacatism seems to express human resourcefulness, intelligence and an orientation toward the spiritual to an extraordinary degree. One may indeed ponder whether the modern world, with its persistent, ambivalent quest for meaning on the one hand, and its problematic issues of illicit drug abuse and addiction on the other, might even be able to learn something from the native Mexicans who have achieved a cultural expression as noteworthy as their practices and teachings surrounding the sacred mushrooms.

Chapter Two
An Account Concerning the Hallucinogenic Mushrooms by Luis Reyes G.

The following text contains an account from Amatlán de los Reyes, a village in the State of Veracruz, in the vicinity of Cordoba. The indigenous people of Amatlán speak Nahuatl, the language of the Aztecs and other Nahua peoples. The account, called *nanakatsitsen* ("little mushroom people"), is presented by Luis Reyes G., a native of Amatlán and "a Nahuatlato (a specialist in Nahuatl)" who collaborated with Wasson by transcribing such accounts from local sources and translating them from Nahuatl into Spanish (Wasson 1980, 32). The theme of the little-people-of-the-sacred mushroom is deeply rooted in Mesoamerican traditions and takes various forms, as illustrated also in some of the other texts offered in the present volume.

The Spanish text underlying the English translation presented here was based on an audio recording of a narrative by a local resident, made by Wasson during a visit to Amatlán in the first week of June, 1960. An interesting detail of this brief article as published was its inclusion of the informant's own words in Nahuatl, along with the Spanish version rendered by Reyes G., a format recalling some of Sahagún's passages dealing with the sacred mushroom. The reader should note that the English text presented below has passed through a double language filter, for it is based upon the Spanish translation, not the Nahuatl original.

Wasson (1980) credits this narrative to Amatlán resident Felipe Reyes F. The comments and anecdotes Wasson gathered from various firsthand sources were often extraordinarily rich in pertinent details. Another account translated by Reyes G. from another local resident, whom Wasson identifies as Rufina de Jesús, referred to a matter of whether the sacred mushrooms would be ground up or not before being taken. For Wasson, this reference supported the idea that grinding the mushrooms was an ancient practice formerly widespread in Mesoamerica, even though it was not observed among the Mazatec and several other groups he studied (1980, 32). This was an important point because the motif of the mushrooms being ground on a *metate* by a maiden established a critical link between a figure represented in one of the Mayan mushroom stones, and extant Mixtec practices that Wasson first witnessed in July, 1960, a month after his visit to Amatlán (see Chapter Four, *The Mixtec in a Comparative Study of the Hallucinogenic Mushroom* by Robert Ravicz).

Apart from its portrayal of the *nanakatsitsen*, perhaps the greater significance of the following account lies in its placement of traditions concerning the sacred mushrooms among present-day speakers of Nahuatl within the State of Veracruz. That such traditions, first written about four centuries ago by chroniclers of the Conquest, should have persisted to the present in spite of vigorous efforts to extirpate them would seem to bear witness to the remarkable importance of the sacred mushroom in native Mexican culture.

*   *   *

Reyes G., Luis. 1970. Una relacion sobre los hongos alucinantes. *Tlalocan* 6:
    140-145.

This text here with which I have been furnished is from a woman of approximately 70 years of age. She is the daughter of an empirical midwife who lived in Amatlán de los Reyes, Veracruz, Mexico, until the first decade of this century. In this place, one observes pre-Hispanic and Christian traits forming part of the religious syncretism that has been present up to the present moment. But in these past years it is tending to disappear in the town mentioned above, the place of residence of the informant.

For the old generations of this place, the land is a living being that "is watching us," "it is listening to us." It provides hunting and good harvests and is worshipped in various ways. The snakes with a "feather headdress" are its symbol, are untouchable, and the name of these animals had better not be spoken in the fields (*kuuatl tipeyolohtle*—snake, heart of the hill). They made offerings of tissue paper, red and yellow or red and black. Also, red cups, flowers, *copal*, tobacco, eggs, bread, and chocolate are carried and candles are left on the mountain or are put in the ground

The earth as mother of every living being is also that which gives knowledge of medicine to humans, knowledge that is revealed in dreams or by "death and returning to life" of the individual.

But apart from the revelation that earth by itself gives to certain individuals, any person can bring on a state of ecstasy by ingesting narcotic mushrooms as a way of consulting the inhabitants of Tlalocan for whatever they wish to know. This includes what could be called medical consultation. These inhabitants of Tlalocan who come to the aid of the narcotized are dead, unbaptized children who are now *xokoyomeh* (beams of light) of blue color who live with the Father and Mother of this mystical place, the entrance of which is in the caves.

Upon bringing about the state of trance, the saints and Christian virgins are invoked. And it is said that in order to obtain better results, the mushrooms should be taken on the holiday of *Yehwatsin* (the transfiguration of God, August 6). Positive declaration of this is based upon the fact that the Biblical passage in which Jesus comes into contact with heavenly beings is commemorated on that date.

The account given below was obtained in the Nahuatl language. For the translation that I have done, I have tried to conserve the syntactic forms used by the informant.

Nanakatsitsen (the little mushroom men)

1. The ancient fathers and mothers had a custom, whenever they would lose something, or even if they just wanted to know something whether it be where their spouses are, or to know who has bewitched them, or whether they would heal or if their sickness would last long. They would take mushrooms that they called *tlakatsitsin* (*hombrecitos*, little men), and these would answer, and tell them whatever they wanted to know.

2. When they have said that they are going to take them, whether it be early or in the afternoon, they go to the field to pick them.

3. It is said that the season of the chili harvest, in August, is when more are found, and then they are picked underneath the chili plants, they are found by two, by three, and if they desire it they go gather them. And it is also said that they answer more if they are taken on the holiday of *Yehwatsin.*

4. If they have already gathered them and now want to take them, they place them in a red cup and put red flowers on top and then they go to the church, virgins go and place them before the saints, they are going to inform the mushrooms so that they may answer, so that they will not upset them because the mushrooms say "you don't respect us, you call us bull excrement," therefore the person who is going to take them prays. If a sick person in bed is going to take them, the mother or father will go to the temple and pray.

5. And when they are going to be taken it is (or may be) necessary that nobody makes any noise; the dogs are going to be tied up faraway. And the cock that they know is going to crow will be put somewhere else faraway; nobody will make noise, nobody will shout.

6. Whoever takes them first says a prayer, requesting that "the medicine" (i.e., the mushrooms) may answer, holding the mushrooms in the smoke of incense (*sahumo a los hongos*). Then he lies down and proceeds to eat them one by one. It is said they would eat twenty, and those with a high tolerance would eat thirty.

7. And a woman will care for whoever is taking them, no one else will be there. It is said that if someone makes a noise or shouts, then he who ate the mushrooms goes crazy. Therefore, if he who ate them is brought to his feet, then he grabs whoever is tending him, he again lies down but doesn't make a sound, he only grabs her.

8. If someone of bad character, if a woman without respect eats them, they don't answer anything to her. She goes on without direction. She feels as though snakes are attacking her. And when something like this goes on without direction she becomes flustered, then they take off her *huipil*. Then they dress her, they dress her in her husband's shirt, they also put his pants on her, all the clothes that were just taken off the man, and she is calmed with them, and once again goes back to sleep.

9. And the person who takes them well and sincerely begins to see many little men like children. He begins to talk with them and these little men tell him everything because Tlalocan is where they are from; so it is the earth that answers because we are standing on it, hence it knows everything. The earth is watching us and it answers.

10. Whoever took the mushrooms, you will then see him seated and he says: "Oh God, hear the dogs ... not making noise. Listen like our neighbor would keep watch over us, he says we are drunk but it is not true."

11. These little men tell him everything he wants to know, if something is lost they will tell who took it, if your spouse goes somewhere, they will tell you where, if someone is annoyed with you, if they talk behind your back, if they dishonor you, you will know everything about it.

12. Also, if you will be rich someday, if somewhere down the road you will remain (or become) poor and you will have nothing, you will know everything there.

13. If you are sick they will tell you how to get better and who will cure you. Along with curing they will also give you a massage.

14. If something inside of you hurts, then with their little hands they will massage you. You feel as though "they settled your stomach." Your stomach and innards will make noise while they are extracting the sickness from you.

15. And if not, you will see that they will open up your stomach. They will pass over you repeatedly, extracting the sickness from you.

16. Women used to take them very often, but not anymore, now they are afraid to.

17. Also, there were certain seeds that they would call "virgin seeds," but these are no more, they are not found anymore along with the little men themselves, although it may be that now you can find them when it rains, in the month of August. (Ed.—this refers to morning glory species whose seeds contain ergot alkaloids. Two such species are used: *Rivea corymbosa*, known as *ololiuhqui* in Nahuatl [*badungás* or *badoh negro* in Zapotec]; and *Ipomoea violacea*, *tlitlitzin* in Nahuatl [*badoh negro* in Zapotec][Wasson 1973]).

Chapter Three
Sacred Mushrooms of the Matlatzinca by Roberto Escalante H. and Antonio López G.

Following is an exclusive study of traditions concerning the sacred mushrooms among a cultural group of rather limited distribution, the Matlatzinca. The Wassons (1957) noted two references to the Matlatzinca in primary literature from the sixteenth to mid-seventeenth century mentioning the sacred mushrooms of Mexico. One of these references, brought to their attention by Robert Weitlaner, was in a report by Gaspar de Covarrubias dating from 1579 on the region of Temazcatlepec, in which it is noted that the native tongue in that locality is Matlatzinca. The report also cites an informant's statement that before the natives became Catholic, "they were wont to pay in tribute [to the Lord of Mexico], whenever they were asked to do so ... and they would give mushrooms on which people get drunk" (Wasson and Wasson 1957, 220).

The use of the sacred mushrooms at the time of contact is also indicated by entries in a Matlatzinca lexicon dated 1642, compiled by Fray Diego Basalenque. Citing this source, the Wassons note: "The inebriating mushroom is the *intza chohui*, and we discover that *chohui* means both 'mushroom' and 'fiesta'!" (1957, 230). The text presented below reveals that the latter term (*chhówi* as Escalante and López render it) has persisted among the Matlatzinca as a general reference for mushrooms. However, the designation of the inebriating mushroom as a type of mushroom (*intza chohui*), as indicated in the lexicon from 1642, contrasts sharply with the situation of the ethnographic present. As described below, among the contemporary Matlatzinca the sacred mushrooms are not even referred to as mushrooms *per se* but rather are recognized in a wholly different category, of things regarded as sacred.

Originally published in Mexico in 1971 by the Museo Nacional de Antropología, Escalante and López' article has been cited accordingly in a number of references, including Wasson (1980), Guzmán (1983, 1990), and Ott and Bigwood (1978) (although the latter gives 1972 as the date). But in contrast to most other literature sources, book or article, in English or Spanish, searches conducted through Inter Library Loan offices at more than a half dozen college and university libraries over a period of years proved unsuccessful in locating this article, at least from the bibliographic information of its original publication.

This impasse was finally resolved with the collegial assistance of Dr. Gastón Guzmán as described in the Preface. But in addition to furnishing the text, the copy he so kindly provided also yielded a further, unexpected piece of highly relevant information. The top of the first page contained a bibliographic citation, apparently not part of the original page as published but rather added by typewritter after the fact, differing from that found in the literature sources where this study had been cited. This typewritten addition seemed to suggest the article was republished in 1972, in Volume 2 of the *Proceedings of the 40th International Congress of Americanists*. Following up on this intriguing new clue, Inter Library Loan was soon able to trace at least one lender with this source. If anyone has tried unsuccessfully to obtain this article previously, they may likely be able to access it now by using this citation, and it is presented here accordingly. I thus offer an additional vote of gratitude to Dr. Guzmán for this extra, unforeseen piece of valuable bibliographic information.

*   *   *

Escalante H., R. and López G., A. 1972. Hongos sagrados de los matlatzincas. *Proceedings (40th International Congress of Americanists)* 2: 243-250.

The Matlatzinca form part of the Otomian linguistic branch, one of the oldest in Mexico, which includes the Otomi language and Mazahua, which is very closely related to Otomi; Pame, which is spoken in San Luis Potosi; Ocuiltec, which is spoken in the state of Mexico; Matlatzinca, which is spoken in the same state, and Chichimec now reduced to five hundred speakers, who are located in Guanajuato.

This branch is very important, because according to archeological and linguistic studies, it is of a very old occupation in our country. As it is also known that the majority are farmers, they are probably related to the pre-Classical cultures.

At present the Matlatzinca are located in a single town, San Francisco Oxtotilpan, Mexico, which is about 39 kilometers southwest of Toluca, along the Temazcaltepec highway (Ed.—a map is provided in the original text, p. 244). The Matlatzinca live in the valley of the Rio Verde, a valley surrounded by forests of mixed type, where there are pines, white oaks (*encinos*), red oaks (*robles*), firs (*oyameles*), etc.

The settlement pattern is dispersed, the population extending along the valley for about four or five kilometers. The corn fields (*milpas*) are in town, and the houses are grouped in nuclei adequately separated from one another. A population of 2,500 inhabitants has been calculated.

The Matlatzinca categorize edible mushrooms or other types in one class, and the hallucinogenic mushrooms in another, as something sacred and not specifically as mushrooms. In the present article we examine this native taxonomy the Matlatzinca have for mushrooms; and the use that is made of them.[1]

In the publication by Cazés (1967) various terms for mushrooms are presented that refer to varieties, and not merely a generic name; which drew the attention of the authors, because the mushrooms are specific to the territory of the Matlatzinca, and there existed the possibility of finding a more extensive taxonomy. Although they are significant neither economically, nor from a nutritional point of view, the way in which they are used is important for anthropology.[2] Towns of Nahuatl speakers close to the Matlatzinca do not have the same number of terms for fungi, and use terms to identify only six species (Schumann and Garcia de León, 1970). Among the Matlatzinca we have to date collected specimens of more than fifty species and their native names, whether they are utilized or not.

This aspect of the meaning is very important because there are certain cultural complexes which require the nomenclature of the native languages; for example, when we consider the entire series of words in Mexican Spanish related to the elaboration of maize, the parts of the plant, and the utensils that are used, we note that all the words are of Nahua origin; we have, for example, the words *nixtamal, atole, tamale, jilote, elote, olote, comal, metate,* etc., showing that this cultural complex has passed from Nahuatl to Spanish.

To designate snow the Eskimo have more than five terms; in Spanish there is one; and in Nahuatl there is no word for snow. But for the Eskimo this is essential because in the place where they live, they must consider different classes of snow; the fresh snow that is falling, the snow that is hardened on the ground, the snow than can be used for building igloos, etc., and therefore there are different words for all these states. These taxonomies, these understandings of the world are directly reflected in the language. In Matlatzinca certain prefixes are used, for example the prefix *chho-* which is put in front of the name of any mushroom (that is to say, except the sacred) the generic term for which is *chhówi*. This prefix contrasts with others that designate human beings, plants, trees, liquids and sacred things.

Among the Mazatec, the general word for mushroom is /*thai³*/ (Wasson and Wasson, 1957; orthographics modified), whereas the hallucinogenic mushrooms are called /*xi³ tho³*/, this last term meaning "that which grows of itself" or "that which is born of itself." To name the different species of hallucinogenic mushrooms they always use the term /*xi³ tho³*/; for example, the mushroom called "*angelito*" or "*pajarito*" (*Psilocybe mexicana* Heim) is /*?nti¹ xi³ tho³ ni⁴ se³⁻⁴*/ which is analyzed as follows:

/*?nti¹*/ = reverential
/*xi³ tho³*/ = that which grows of itself
/*ni⁴ se³⁻⁴*/ = bird (*pájaro*)

This contrasts with the term used for *Schizophyllum commune* (Ed.—the species cited here is a common, widely distributed polypore, one of many sometimes referred to as bracket fungi or conchs, which have tough tissues and corresponding-ingly persistent fruiting bodies that do not disintegrate readily; among these *S. commune* has a unique spore-forming surface often described as "split gills," the literal translation of the genus name *Schizophyllum*) also called *"pajarito"* which is /thai³ ni⁴ se³⁻⁴/ and is analyzed:

/thai³/ = mushroom
/ni⁴ se³⁻⁴/ = pájaro

In the Matlatzinca language the word to designate mushroom is /chhówi/, and the word for hallucinogenic mushroom or *"santitos"*[3] is /netochutáta/, which is broken down into four elements:

/ne-/ = plural
/to-/ = reverential, diminutive
/táta/ = Mr.; grandfather, ancestor

In this language there are prefixes that classify, for example, /sa-/ which is used thus in names of trees; the names of herbs carry the prefix /xi-/; the names of liquids carry the prefix /chi-/; the names of human beings /we-/; and "mush-rooms" have the prefix /chho-/; the syllable /chu-/ denotes sacred things, espe-cially sacred mushrooms. Here is seen the principle of taxonomy that this lan-guage uses to classify the phenomena and things of the world; they consider that the trees are different from the herbaceous plants, and the herbaceous plants different from the "mushrooms," and they are different from man, and from sa-cred things.

In contrast to the names of "mushrooms" formed with the prefix *chho-*, they distinguish things and phenomena they consider sacred with the prefix /chu-/. They consider as sacred the sun, the moon, also the clouds, the rain and light-ning (*el rayo*), which they call /nechutáta/ or *"santos;"* in the Matlatzinca lan-guage one says "the saints are coming" (i.e., the waters), and also "the saints (= clouds) are roaring, they walk rumbling" (Text 1, e). Also by connection or as-sociation with the clouds, or with this sacred concept, they call Nevado de Toluca (a volcano, with national park status) "the house of the saints," /nromani nechutáta/; and fire is /chutá?/ because they consider it sacred. They also con-sider as sacred the church, and the images that are inside. The saints and the clouds are designated with the same word /nechutáta/; the sun and Christ are both designated with the word /nchutáta/; the Virgin and the moon are desig-nated with the term /nchunéne/, and also the *Psilocybe* takes the prefix /chu-/.

In reference to hallucinogenic fungi, they speak of *Psilocybe* not as mush-rooms but rather as *"santitos"* /netochutáta/; they say: "the *santitos* teach many things" (Text 1, r).

These sacred things are related with nonordinary or extraordinary experi-ences, for example the mushrooms that produce hallucinations. There is a belief that a person who is struck by lightning and does not die acquires powers to prophesy, to cure, and knowledge of medicinal herbs. In Nevado de Toluca everything relating to the crops, the animals, etc., is predicted annually.

The hallucinogenic mushrooms (*Psilocybe muliercula* Singer & Smith) are called in Spanish *"santitos"* or also *"honguitos de San Juan,"* because, according to them, they begin to sprout June 24. The season in which they appear is generally calculated to be between June and August. The end of August is when the specimens placed in the National School of Biological Sciences of IPN were collected, and they were the last ones in the region.

The habitat of *Psilocybe muliercula* has been described already by Dr. Guzmán (1958) at relatively high altitudes, in a ravine that produces a clearing in the pine forest. In a more recent article (Guzmán and López, 1970) a new locality is reported, San Francisco Oxtotilpan, presenting a different habitat at a lower altitude in a river valley. The *Psilocybe* sprouts from the thick banks of the river, close to a spring with a sown field nearby where livestock are taken to graze. These thick banks have a certain shade in the day, but during good weather they are exposed to sun.

When the *santitos* are collected a flower is left to signify respect and to favor their return the following year. The mushrooms are consumed fresh, they are not to be eaten dried because the *"santitos"* (the supernaturals who are seen) get mad and can punish the person or produce in him a negative experience. Also they are eaten raw not cooked, and without cleaning, even if they have soil adhering to them. They are consumed in one, three, five, or multiples of five, passing them over the smoke of an incense burner with embers and resin of *copal*, accompanied by sugar or fruit. There are no indications of sexual abstinence nor of fasting as exist in the Mazatec territory, it is said they may be eaten an hour after taking food. There is a prohibition on ingesting them with *pulque* or distilled liquor; neither can cigarettes be smoked when they are being ingested or during the experience.

Formerly this *Psilocybe* comprised part of a healing ceremony, during which the specialists or *curanderos* would consume the mushroom along with the patient; and the mushroom diagnosed the illness and prescribed what medicines should be administered. At present this type of healing is no longer practiced. Now people take it out of an interest in having nonordinary experiences, and its use is not very extensive in the population, they fear what they are going to see, what is going to happen; and only the young people take them.

Concerning the effects of the *Psilocybe* and other "magic plants," Drs. Rinkel and Schultes (1965) tell us that "all have in common the following characteristics: they produce a certain dream-like state, marked by an extreme alteration in experience, in the apprehension of reality, in the perception of time and space, and in consciousness of the self. They induce visual hallucination, often with kaleidoscopic motion and usually in rich and brilliant colors, indescribable, accompanied by auditory hallucination and other types, and a variety of synesthesia."

In another study concerning the use of hallucinogenic plants among the Yaqui (Castañeda, 1969) in reference to the "apprehension of reality" we are told that the designation of "states of nonordinary reality" refers to "a strange reality opposite to the ordinary reality of everyday life. The distinction is based on the inherent meaning of these states of nonordinary reality." In the context of Yaqui understanding "they are considered as real, although their reality is different from the ordinary reality."

The type of experience, or hallucinations, etc., that the Matlatzinca have is not the same as that which a person from the city could have, for it is culturally conditioned. They say that the *santitos* are "phantasms," like ghosts (*son "afantasmados," como fantasmas*), because they allow for the existence of things that do not exist or that we cannot ordinarily perceive. They also say that the *santitos* are little people, men and women, who announce themselves when the *Psilocybe* is taken (see Text 1, o; Text 2, v, w). These *santitos*, they say, speak Spanish or Matlatzinca, although their communication has nothing to do with the language. They say that "they teach," that is to say they show pretty things such as flowers, stars; or ugly things like snakes or blood, especially to people who do not lead a very exemplary life, or who take *pulque* or are evil (see Text 1, Text 3b).

Formerly both the *curandero* and the patient ingested the mushrooms; the patient had to lie down and cover himself with a blanket under which smoke of *copal* was introduced, and held, and it had to be done in silence. There was nothing resembling the ceremony of the Mazatec, where there is active participation of all the people present in the session. Among the Matlatzinca nowadays, the experience is a personal thing, each one takes his mushrooms, and when he feels the effects he covers himself with his blanket, lies down somewhere, and sees what the *santitos* communicate to him. The people who are in the house participate in a passive way; they maintain absolute silence and great respect; no one may speak or laugh; the children are sent out of the house. After the experience people ask whoever took the mushrooms what the *santitos* taught him, that is, what he saw or what they did, and what they showed him.

Also the *"santitos"* when they speak give prophecies as to what is going to happen. In one of the texts presented, the death of a gentleman who took the *"santitos"* is predicted. It says also that they cure, they speak about the disease that the patient has, that is they diagnose; also they reveal medicines through the type of images that are seen; and it is also said that they give massage. This is when the person conducts himself well; when one behaves badly, then the *"santitos"* punish, "they flog," they appear with their lash, their whip; it is said that they make one cry. Probably because of this the people have respect for them and do not use them indiscriminately, but there are many who prefer not to consume them.

It is interesting to observe in the texts that out of all the stories obtained none refer to the personal experiences of the narrator, because with the reduced number of people who have ingested the fungus it is difficult to find any person who can communicate the direct experience, in addition to which they are not willing to confess to a stranger as one wishes the use they make of *Psilocybe*.

Translations of texts from the Matlatzinca

Text 1

a. (The) deceased, my dad (*papá*), told me (that) he ate those *"santitos."* b. He told me (that) they taught him c. many things. d. Later he told me that he heard e. that the clouds were thundering, (that) they walked rumbling; f. (that) what was happening to him was pleasant. h. Later of that (of which) he ate, i. he told me, j. that they taught him when he was a soldier; k. when they touched his bugles. m. Later they taught him some flowers (*Después le enseñaron unas flores*—this is the idiom mentioned above, that visionary images seen under the effect of the mushroom are said to be "taught to" the percipient, Ed.), n. which were pretty. o. Later those *"santitos,"* those little mushrooms (*honguitos*) were thus (like) little gentlemen, p. (when they were) small mushrooms. q. Later he told me (the) deceased, my dad, he told me, r. (that) those things the *santitos* taught him. s. They told him: t. "When you went to fight, *Diosito* did not allow you to die as yet, n. (when) you fought thus." v. Then he said; w. (that) they then told him, thus: "You are no longer going to wait many years, x. because you are going to die." y. Then my dad, (the) deceased, said he was frightened. z. "Neither way (if) I am going to die soon, only God knows." a'. Then we remained in town; b'. soon he no longer ate much, he began to become ill.

Text 2

a. Nowadays, b. we believe (that) they truly are *"santitos."* c. Because a gentleman, d. told me so. e. He said f. (that) when the *maicitos* (Ed.—literally "little corns," referring to immature developmental stages of ants in the genus *Liometopum*, more popularly known as *escamole*, an insect delicacy of native Mexican cuisine [see Conconi, 1982]) are broken, g. (that) their juices (*juguitos*) spill, h. but, i. (that in reality) it is their blood that spills. j. Later (to) that gentlemen, k. I encountered him again l. (and) he was crying. m. He was carrying his things (*yuntas*); n. already leaving, going to his house. o. He was crying, p. so I asked him, q. "Why are you crying Mr. Aureliano?" t. And he, u. did not answer me. v. Then (he said) that those *"santitos"* were teaching him some people; w. sometimes gentlemen, sometimes ladies, x. (that) they were telling him, they were telling him, that they are *"santitos"* because they teach us so many things. y. (Because) they do not kill, z. they do not poison.

Text 3

a. They also taught my sister; b. they taught her enough flowers, (in) a tiny garden (*jardincito*) (which) was pretty. c. Then (she) got mad (with) our dead dad; d. and also she was very bad. e. (Therefore) they administered the *"santitos"* to her; f. she was shouting (that) they hit her, because my sister thus got mad a lot. g. Then they told her (that) they were going to take her away, h. but they just beat her, (and then) they left.

Text 4

    a. (For) the first time Lourdes ate the *"santitos,"* *b.* he said that they accounted him, *c.* enough things. *d.* Later he said they taught him a tiny lagoon (*lagunita*), *e.* (where) there were many tiny boys (*niñitos*) dancing. *f.* Later he said he saw (some) flowers; *g.* that he had never seen.

Text 5

    a. The *"santitos"* are born near the edge of the river. *b.* Now already if there are enough. *c.* We are going to look for them, *d.* I hope that we find them. *e.* Now already if we arrive at the house, *l.* we are going to hold them in the smoke of the incense burner. *m.* Now we are going to eat them with sugar. *n.* You eat no more than four. *o.* Now we are going to lie down, *p.* so that they work well; *q.* but, they are going to get mad, *r.* and they are not going to work. *s.* Now we are going to put *copal* under his blanket. *t.* They have left so that they do not make any noise. *u.* What did they teach you? *v.* They taught me this: because I grumbled to my mother, *w.* they hit me with challenge enough; *x.* they beat me (with) a quarter of the horses. *y.* And they told me: "Do not do that way unto your mother anymore, *z.* if we are not going to take hold of you again." *a'.* I am no longer going to do thus to my mother.

Text 6

    a. (Thus) the deceased, my grandpa, said to me. *b.* He told me a time ago, *c.* (that) he once ate them. *d.* He told (me) that he saw them, *e.* that they were like the little kings (*reyecitos*) in a deck of cards, (that they were) like dolls. *f.* He said that they were very small.

Notes

    1. C. Lévy-Strauss, 1962, *The Savage Mind* (*El Pensamiento Salvaje*, 1970, Breviario, FCE, México). "Native classifications are not only methodical and based on carefully built up theoretical knowledge. They are also at times comparable from a formal point of view, to those still in use in zoology and botany" (1962, 43).
    2. A. H. Yunatov, (USSR), *Utilization of wild flora by nomadic populations of central Asia*, VII Congreso Internacional de Ciencias Antropológicas y Etnológicas, Moscú, 1964, Tomo 5, "the discovery of useful wild flora employed by local population in their economy and everyday life is a common task of both the ethnographer and botanist. The efficiency of such an integrate approach has not yet been sufficiently realized though it promises interesting possibilities. In many cases an ethnographer, resorting to the qualified assistance of a botanist, can verify the information collected by questioning, and give rational explanations concerning the assortment of the plants used. On the other hand, ethnographical data provide the botanist with good orientation in their search for the most valuable and promising species of useful wild plants. This is especially important in assimilating the experience of popular medicine, in employing pasture resources, in finding technical vegetative raw materials."

3. *Psilocybe muliercula* Singer & Smith = *Psilocybe wassonii* Heim.

Bibliography

Castañeda, C., 1968. *The Teachings of Don Juan: A Yaqui Way of Knowledge.* New York: Ballantine.

Cazés, D. 1967. El pueblo matlatzinca de San Francisco Oxtotilpan y su lengua. Acta Antropológica, Epoca 2a, Vol. III, No. 2, ENAH, Soc. de Alumnos.

Guzman, G. 1958. El habitat de *Psilocybe muliercula* Singer. Revista de la Sociedad Mexicana de Historia Natural 19: 215-229.

———, and A. Lopez G. 1970. Nuevo habitat y datos etnomicológicos de *Psilocybe muliercula.* Boletin de la Sociedad Mexicana de Micología 4: 44-48.

Lévy-Strauss, C. 1962. *The Savage Mind.* Chicago: University of Chicago.

———. 1970. *El Pensamiento Salvaje.* Breviario 173, FCE, México.

Rinkel, M., and R. E. Schultes. 1965. Transcultural Significance of "Magic Plants." Cambridge, MA.

Schumann, O., and A. García de León. 1970. Cartilla de Lengua Mexicana. Colección "Caracol" No. 1, México.

Wasson, R. G., and V. P. Wasson. 1957. *Mushrooms, Russia and History.* New York: Pantheon Books.

Yunatov, A. H. 1964. Utilization of wild flora by nomadic population of Central Asia, VII. Congreso Internacional de Ciencias Antropológicas y Etnológicas, Moscú.

Chapter Four
The Mixtec in a Comparative Study of the Hallucinogenic Mushroom by Robert Ravicz

Following is an anthropological report on the mushroom ritual among the high-land Mixtec in Oaxaca, based on participant observation by Ravicz, who documented it in 1960 along with R. Gordon Wasson. The Mixtec inhabit a region occupying a large part of northwestern Oaxaca known as the Mixteca.

Ravicz was noted for ethnomycological observations among the Mixtec by Heim and Wasson in 1958. By then, he had already observed that the mushrooms were used in Zacatepec in the district of Putla, and in villages of Juxtlahuaca such as San Pedro Chayuko, Agua Fría, and Santa María. He also noted the tradition Wasson first documented among the Mazatec, that whatever is said under the influence of the mushrooms, "it is the mushroom that speaks and not the person who digested the mushroom" (*c'est la champignon qui parle et non pas la personne qui a absorbé le champignon*) (Heim and Wasson 1958, 100).

Ravicz' work since these early observations has proven invaluable for understanding not only the Mixtec, but also the greater cultural context of the sacred mushroom in Mesoamerica, especially as relates to the ancient civilization of the highland Maya. One of the Wasson team's many successes has been the interpretation of the "mushroom stones" of Guatemala as evidence of mushroom cultism during the Classic and pre-Classic stages of Mayan culture. The first description of such an artifact was that of Carl Sapper in 1898, who with admirable restraint avoided jumping to the conclusion that it specifically depicted a mushroom *per se*, and merely characterized it as mushroom-shaped. The specimen Sapper described was a particularly striking example which has been referred to since as the Reitberg stone, after the Museum in Zurich where it has been displayed (Wasson 1963, 52).

In the 1950's, as V. P. and R. G. Wasson penetrated firsthand into the mysteries of the sacred mushrooms of Mexico, they realized there had to be a connection between the Mayan mushroom stones, and the practices they observed among contemporary groups such as the Mazatec, and they were the first to make such a suggestion (1957). As they discovered, archeologists who knew of the mushroom stones were unacquainted with the early reports on *teonanácatl,* or with the research of Schultes. By that time, more than a hundred such artifacts had been identified, but there was a problem. How exactly did these ancient Mayan mushroom stones figure in relation to nanacatism lingering in present day Mexico? The rituals of the sacred mushroom reported from contemporary Mesoamerica involved various ceremonial items, but nothing like the Mayan stone mushroom artifacts appeared among them. Even more problematic, based on field investigations by the Wassons in 1953, contemporary Mayan peoples seemed to have no knowledge of the sacred mushroom. Isolated reports from within two Mayan groups of southern Mexico, the Chol and the Lacandones, of individuals using *Psilocybe cubensis* ceremonially for curing or divination have been cited by anthropologist Peter Furst (1976, 75). Wasson emphasizes that such reports have remained unsubstantiated, and that analysis of the contemporary cultural context has been complicated in recent decades by "the activities of trippers and hippies" (1980, 181).

References to "mushrooms that intoxicate" and similar such phrases have been found in handwritten Mayan (Cakchiquel) word lists and dictionaries compiled several centuries ago, strongly indicating an awareness of the sacred mushrooms among the Maya in historic times. But the regions of Mesoamerica where mushroom ritualism has lingered into modern times are all in Mexico, not Guatemala. The Wassons and others noted a gap in time, cultural-geographic space, and tradition. Some kind of evidence was needed to help secure the mushroom connection between the extant Mexican rituals and the ancient Mayan artifacts.

> Certainly, we have not discovered the tie that would unequivocally bind the mushroom stones of Guatemala and Chiapas with today's intimate folk cult. That such a link will be discovered is unlikely but not impossible. Perhaps on some holy hill or *cerro,* or deep in the recesses of some cave, in a remote corner of Oaxaca or Chiapas, the Indians still direct their humble supplications to a stone image of a mushroom, unbeknownst to circumambulating ethnologists. Unless someone discovers such a survival, how can we hope to establish a connection between the ancient stone carvings and the divinatory mushroom? (Wasson and Wasson 1957, 285)

One mushroom stone in particular, which the Wassons referred to as the Namuth artifact, displayed on its stipe a female figurine bending or kneeling over a flat, sloping surface. In the late 1950's, when the first classic works of the Wasson team were published, the interpretation of this figurine was enigmatic. As luck would have it, the Mixtec ritual described here by Ravicz provided an unexpected insight, helping to solve the riddle of the Namuth stone. And that solution, in turn, helped establish a more specific connection between practices of contemporary native Oaxaca and the Mayan past.

In July 1960, Wasson and Ravicz attended a Mixtec *velada* in the village of Juxtlahauca, Oaxaca. The article presented below is based on Ravicz' observations from that ceremony. By this time Wasson had already published his extensive findings from several years of field research among various native Mexican peoples, especially his experiences among the Mazatec such as in 1955, when he became a full-fledged initiate in the rite, eating the mushroom and experiencing its effects along with the *curandera* and others in two successive *veladas*. But in the Mixtec ritual, Ravicz and Wasson documented something neither had ever seen before, which shed an unexpected light on the riddle of the kneeling female figurine of the Namuth stone. As Wasson states:

> The notable feature of this *velada* was the preparation of the entheogenic mushrooms: Juventina, the youngest daughter of the family, a child, a maiden (*doncella*), ground them on a *metate* over which she leaned ... It was clear to me at once that the *doncella* was performing the religious office of the female figure in the Namuth artifact. There could be no doubt about it ... Juventina in Juxtlahuaca was, all unknowing, the key to the highland Mayan *doncella* some 2000 years before (Wasson 1980, 179-181).

If indeed a common cultural pattern connects the ancient Maya with Mixtec tradition, as the Namuth stone and the ritual in Juxtlahauca tend to confirm, three possible types of explanation come to mind. The mushroom motif of the *Metate* Grinding Maiden—it is too easy to imagine a character by such a name in some fanciful, lost myth of origins about nanacatism in Mesoamerica—may have been invented independently in Guatemala and Oaxaca with no external influences, in which case its occurrence in these two places and times would be sheer coincidence. Or it may have originated in one place and then diffused to the other by normal processes of cultural contact and exchange, which seems less unlikely.

But more compelling than either of these alternatives is the idea that both cultures may have inherited this interesting motif from the shared stock of their distant ancestry, a case of common cultural descent and divergence, with retention of this ancient feature in both lines, and diffusion a secondary influence. Such a framework can account for differences as well as similarities, thus fitting well with the facts of the Mesoamerican cultural mosaic we observe in reality.

But if this is indeed an instance of homology in cultural evolution, it would indicate that mushroom usage is truly ancient in Mesoamerica, and must have already been present before the rise of archeologically known Mayan and Mexican civilizations. Such a perspective, however hazy, is not at all inconsistent with well framed, widely held views that the use of hallucinogenic plants and fungi in shamanistic practices of the New World is the cultural cousin of ritual use of the fly agaric (*Amanita muscaria*) by Siberian shamans. The Mayan mushroom stones enter the archeological record close to the time when stone carving first appeared, so the absence of similar artifacts of even greater antiquity does not stand as evidence against the hypothesis of an even more ancient use of mushrooms in the region.

As though to substantiate the link between the mushroom stones of the Maya and the Mixtec ritual, a set of artifacts from an archeological site in Guatemala soon helped further establish a connection between stone grinding implements and the mushroom stones: "nine miniature mushroom stones with nine miniature *metates* and *manos* ... linked by physical association" (Wasson 1980, 181) were found at the Mayan site of Kaminaljuyu. In the discussion below Ravicz cites this discovery, a recent one as of the early 1960's when his paper was published (see Borhegyi 1961). Nowhere in the literature on extant Mexican mushroom ceremonies have any ritual implements such as the mushroom stones ever been reported, so just how these intriguing sculptures figured in Mayan tradition remains unclear. One may wonder whether they stood as images placed upon an altar for the ceremony, somewhat as pictures of saints are nowadays used in Oaxaca.

Ravicz' importance in Mesoamerican ethnomycology deserves recognition. Not only is his discussion of the subject astute, but his interest in it was serious enough that for the ceremony described here, held the night of July 5-6, 1960, he ingested the ground mushroom preparation, along with Wasson and the *curandera*, willing to withstand its effects in order to gain a better understanding. Also, Ravicz facilitated the visit to Juxtlahauca and arrangements for the ritual through his personal acquaintances there, especially Guadelupe González Vega, who provided the investigators with lodging in his home, and whose aunt served as the *curandera* for the ceremony. Wasson and Ravicz studied together in the Mixteca again in 1961, with Roger Heim also in attendance. Dr. Margaret Houston lent invaluable assistance with the translation of this piece and offered comments to aid its clarification, many of which form the basis of parenthetic editorial notes.

*   *   *

Ravicz, Robert. 1960 (1961). La mixteca en el estudio comparative del hongo
  alucinante. *Anales del Instituto Nacional de Antropología e Historia* 13: 73-
  92.

1. Introduction

Among the points of greater interest in recent years for Mesoamerican studies is the one pertaining to the hallucinogenic mushroom (the *teonanácatl* of the Aztecs).[1] The content of this work touches upon several branches of anthropology, with regard to the physiological and psychological effects in the human organism, or with respect to what archeologists and ethnologists can say about the meaning of the fungus within pre-Columbian cultures and in the present.

By itself, the phenomenon of hallucination seems to be inherent to the human condition, so that one always has to take into account that its manifestation varies according to the values and demands from one culture to another. Accordingly, we have discovered that it has played an important role in many societies, for example in the relationship of the religious life with the social structure, the economy, medicine and many other aspects.

The dreams and visions are manifested and realized in distinct ways, and at the present time have a very ample worldwide distribution. The main objective ought to be to seek and find explanations about the unknown context.

The hallucinogenic mushroom is one of the agents that actually permits us to pursue that goal. It is employed ritually in Mexico, and perhaps in Central America, but the distribution and use of the various species covers a considerable part of the whole world.[2] Fortunately, Mexican history and ethnology provide the best conditions known so far for securing the study of various cultural problems, considering the mushroom as a focus to better illuminate the life of pre-Columbian Mexico or to examine cultural change. The data presented give an idea of the uses of the mushroom[3] today and the linguistic and geographic range in which it is used.[4]

II. The Mixtec[5]

Cultural Framework

One of the purposes of this study is to present some data on the fungus among the Mixtec that until now have not been published.[6] Another purpose lies in making a study of comparison between the data already known.

Of the three regions into which the Mixteca is commonly divided, the Alta, the Baja, and the Costa, we will make reference only to the first two because the fungus usually does not occur at altitudes below 600 meters.[7] (Ed.: The Alta Mixteca is the higher part of the Mixtec highlands; the Baja is the lower altitude highlands, partly in Oaxaca and partly to the west in Guerrero.) In these regions there live more than two hundred thousand people who speak Mixteco.[8] Cultivation of maize provides the nutritional foundation of life, in combination with beans, chili peppers, squash, some fruits and the gathering of a few wild edibles. Meat of hen or turkey is eaten on the majority of ritual occasions, such as those of the *mayordomías*. Rarely does one encounter pork or beef. This region is located where the Sierra Madre del Sur and the Sierra de Oaxaca come together. In this region there are great differences in terms of altitude,[9] climate and features of the landscape; it is a rugged zone, broken up, where one notices appreciable differences from one small valley to another.[10]

These geographic characteristics have a parallel in the cultural situation. There is considerable diversity in certain aspects of the general culture, in the dialects, the regional and local attire, industries and crafts with their markets. There is great isolation of various places in the Mixteca, so that foreign influences are expressed in unequal degrees from town to town.

Despite the existing differences, there appear certain regularities. The Mixtec are a peasant folk, however large or small the community. Due to the way the fields are divided up, a Mixteco must sometimes walk one or two hours to arrive at his cornfield. In general, lands are communal or small holdings, their control not governed primarily by parentage or kinship.

The residential pattern demonstrates a tendency towards patrilocality. Whenever possible, the married sons of a father live next to his house. Each one maintains his own house separately, but it often happens that the women get together to do kitchen work. The kinship system is bilateral. Mutual aid is common between neighbors, relatives, *compadres* (godparents) and friends at harvest time, and when they undertake the construction of a house. The payment consists of a meal which includes meat.

Authority in the family is distributed between the man and the woman. Although the child who wants to marry can express some preference, the decision is often left in the hands of the parents. Civil marriage is more common than a church ceremony, but there are also many cases in which the pair "have only been united." The important thing is the acceptance of the community, which is registered by a celebration, that is, with customary ritual. (Ed.: In other words, neither a church nor a civil ceremony, but a traditional feast to formally present the couple to the community.)

The principal unit is the community, that is to say, the place where you find the "authority" such as the image of the patron saint of the locale. The civil form of authority of the town is the municipality. The religion is administered through the medium of an organization dedicated to maintaining the system of the brotherhoods (*cofradias*). The two organizations (i.e., civil and religious—Ed.) are strongly interlaced, by means of the roster (*escafon*, "big ladder") of degrees and people who provide service. There are many towns where "the elders" (men who have passed through all the ranks of the civil and religious hierarchy—Ed.) dominate various aspects of the political and religious life. They are treated with much respect in the whole town, in view of their experience and wisdom. Another source of respect in Mixtec society originates in the system of *compadrazgo* (godparenthood), which plays a role of great importance.

In part of everyday life, or in relations with people or the saints, the individual lets himself be guided by Catholic rules. In other parts, he turns to different resources, such as magic, that allow him to make contact with a wider universe. There is a complex of beliefs and acts, not Catholic, but well marked off,[11] whose foundation consists of the concept of an animistic universe which is expressed through the various classes of *"dueños"* of lightning, rain, earth and animals, and these reward or punish the acts of man with signs that are manifested in the body, the spirit, or the surroundings. In this manner the beliefs and ceremonies that reflect the Mixtec supernatural can simultaneously explain the relation between witchcraft (*brujería*) and sickness, holding the explanation for the causes, and the methods of treatment. In order to know these, one must precisely put oneself into contact with the extrahuman world, which extends much further than the surroundings and ordinary relations, requiring that the individual must become free of the profane world in order to associate with the other. This is not within reach of the people via common means of divination, although they are of great importance in life. What is really required is the use of an extraordinary means, and the hallucinogenic mushroom is one of these.[12]

The Mushroom

The mushroom has the quality of being animate. There is a conversation between the mushroom and the one who takes it.[13] The fungus knows much and thus is able to foretell. As the mushroom itself foretells, the proposition of ingesting it, as an intention, is what puts one into contact with the spirit of the mushroom. Insofar as the mushroom knows of deeds and activities that man cannot know without its aid, the fungus represents the extrahuman world. In itself it contains a supernatural or sacred force[14] that is related with shamanic wisdom (*la sabiduría*), but this is not the entire essence of the mushroom, that is to say, the being and the power of the mushroom are separate properties. Thus, although it has not been totally personified, neither is it considered as a mere impersonal object.[15] The attitude toward the mushroom is reverent. The *curandera* bows her head before the fungus and kisses it, or the box holding it.

Accompanying this deferential attitude is a great confidence in the efficacy of the cures or prophecies of the mushroom. The latter indicates the cause, the development, and the cure of illness. It predicts whether one will get better, or die of the sickness. The mushroom in itself is curative. They consult it to predict the future, in order to see if a trip that is planned will turn out well, or if the individual will be bewitched (*a embrujar*). If one asks whether one is going to be rich or poor, and if what the mushroom indicates does not come to pass, the individual takes the blame, not the mushroom. If something is lost, the mushroom says where it can be found.

The circumstances for ingesting the mushroom are fixed.[16] The time indicated is the silence of night, and the designated place is the interior of the house. Everything must be quiet when one takes it so that the mushroom "speaks."[17] "If there is noise, such as an animal or child might make,[18] it will not give results, or else ugly things will be seen." "If one is not of great faith, the mushroom, being very delicate, does not speak, it only intoxicates." There are certain days that are good to take it, such as the day of the town Saint. Another person must always be present in order to serve as what they call "the caretaker." The task of the companion is to look after the person who takes the mushroom. If the individual leaves, the companion must follow to see and hear what he says, and to guard him. The companion must listen to the words of the one who has taken the mushroom in order to repeat them to the individual when he returns to himself, at which time they go over everything that has transpired. He who has taken them always remembers everything that has happened to him. The companion can be of any social status, a relative or not, without distinction of sex or age. It seems that one of the effects of the mushroom is such that once eaten, it often makes people leave or run. The individual who finds himself in this peculiar state[19] can regret having left or harm someone else, such as if he encounters an adversary.

Also there is a limit as to the number of people who can be present at the event.[20] This depends principally on the desires of whoever is going to take the mushrooms. More than one person can take them. One can employ the services of a *curandero*, but it is not obligatory.[21] No doubt, the one who wants to know something is the one who personally ought to take them. The presence of other people is necessary,[22] but not sufficient, for there are many other conditions that must be fulfilled.

The mushroom[23] occurs in the rainy season, that is, "it is born with the first rains." It only occurs in certain places, so that one town in the region can have the mushroom at hand and another not, even though they may be close to each other.[24] In addition, they don't occur every year if the climatic conditions are not absolutely favorable.[25]

By knowing how to preserve it, the power of the mushroom can be retained for several months. It is possible to be preserved in a simple way, by putting it for a brief while in the heat of the sun or a lamp. This way it retains its properties for up to six months, a process that has the advantage of facilitating its transport to any other region.

Sometimes they sell it on the day of market, but it is not exposed to sight. Since the demand always surpasses the supply, there being many *curanderos* and people who wish to obtain it, most of the mushroom harvest never makes it to the market, but is distributed through other means, such as through the *curandero* or as a favor to a *compadre* or other person one knows.

The one who collects the mushroom must be a virgin, who should also be a relative of whoever is going to use it. There is no special day or hour to gather it, nor is the act accompanied by any other ritual activities. The same girl, or another youngster whom man has not known, participates in the preparation of the mushroom prior to its ingestion.

Because the mushroom speaks and its words come to pass, still other conditions are required. They are divided into three phases: before taking the mushroom, after taking it, and moments in between including while taking it.

Prohibitions include sexual relations and eating. From one to eight days before as well as after the ceremony, one must refrain from sexual contact.[26] The mushroom must be taken on a fast, which means no eating during the eight hours beforehand. The following day, in the morning, one drinks an infusion of orange leaves, and the chocolate, but one should not eat until midday.

The procedure for conducting the rite is more complicated. It is necessary to arrange a table that will serve as an altar, a task that can be done by any relative,[27] the *curandero* or the companion. After having covered the altar with a clean table cloth, one sets up the other elements in the required arrangement [28]

The inventory for the ritual includes the following:

Tables	A wax candle
Flower vases	The mushrooms
Flowers	Cigarettes
Candles	Alcohol
Glasses	A table cloth
Images	A *petate*
An incense burner	A *mano* and *metate*
Copal	

(Ed.—Fig. 2 is a line diagram of the *mesa* or altar seen from above, showing the relative size, shape and position of items listed above. It is captioned: Arrangement of the ritual objects on the Mixtec altar, before grinding and ingesting the mushroom. Items represented in Fig. 2 are listed by letter, A-M, as follows.)

A. wooden table

B. metal bucket with paper flowers (*margaritas*)

C. white and red candles

D. vases with flowers

E. Virgin of Remedies (Ed.: the virgin of the cathedral in Oaxaca)

F. Virgin of Guadelupe

G. seven paper flowers, four rose color and three white

H. incense burner and copal

I. wax candle

J. box with mushrooms

K. box of cigarettes

L. bottle of *aguardiente*, and another of alcohol (Ed.—shown in the drawing and identified with the letter L, but appearing in the list of the original text with a second letter K, the list containing no L; apparently a typographic error.)

M. clean table cloth

Then the process of the preparation of the mushroom follows. The girl who gathered it, or some youngster of the family of the one who is going to take it, prepares it under the direction of the *curandero*.[29] First one must wash the *metate*, after which the *curandero* prays in front of the mushroom.[30] They are lifted and placed next to the *metate*.

(Ed.—Two photographs credited to Wasson appear in the original publication. One shows the ceremonial altar with its arrangement of ritual props as described above. The other, as indicated by its caption: "The girl grinding the mushroom: her mother directs her. Meanwhile the *curandera* holds the box with the mushrooms in her left hand and a lighted candle in her right.")

He makes the sign of the cross over the open box containing the mushrooms, and passes the necessary quantity to the girl to be ground with water[31] on the *metate*.[32] The *curandero* lights the candle and directs the girl in detail, indicating at each pass how to do the grinding and add water to the mushrooms while grinding, telling her when she is finished with the task. Then very carefully the *curandero* pours the water of the mushrooms into a glass, scraping the surface of the *metate* with his fingers in order not to lose any of the liquid. If some drops fall to the ground, the *curandero* advises that no one touch the wet spot before he performs a "*limpia*" ("cleansing")[33] which consist of repeatedly passing the incense burner over the spot and its surroundings, spreading the *copal* smoke over the place and its outline.

After putting the solution of the fungus on the altar, he places a *petate* before the altar and "cleanses" it after also "cleansing" the ground upon which he is going to put it. Kneeling on the *petate*, which is in front of the altar, the *curandero* prays silently and briefly, continuing this phase with another "cleansing" underneath and above the altar, and then another one of the person who is going to consult the mushroom.[34] With the preparations completed, the *curandero* asks whoever is going to take it to come close to the mushroom, to discuss with it the proposition of consulting it, realizing the one who drinks the liquid is the one to whom it speaks. Once it is consumed, the individual briefly prays and lies down. If all the preparations have been well made, then the fungus will speak.[35]

III. Comparison of the Data

The Fundamental Characteristics

In the following sections we will compare the data relating to the hallucinogenic mushroom of the Mixtec with those previously known referring to other regions.[36] The purpose will be to formulate a Mixtec pattern and indicate to what extent the same thing is encountered in the others. Thus clarifying the problem, we will have a better idea of the general extent and regional characteristics which will lead us to define whether a general pattern exists.[37]

In Huautla there are several words designating the mushrooms. There is one in particular that distinguishes the hallucinogenic mushrooms from common kinds, and which means something like, "that which is borne of itself." In total there are four species, each one with a Mazatec name, such as in Spanish signifies "little angels" and means one of them. Also they are called "blood of Christ."

The mushroom itself is considered very delicate or dangerous, on which account you should not speak to anyone about them. They kill anyone who eats them in a ritually impure state. In order to be pure, they must refrain from sexual relations during five days before and after eating it. If the person who collects them in order to give them to another is not pure, the mushroom can kill the one who eats it, or drive him crazy. One or more of the same family can eat it. Generally, it is not the sick person nor his family who eats the mushroom, but the *curandero*, who relays to them in a high voice what the mushroom shows him, sometimes by means of a rhythmic song. They are taken more or less at nine in the evening, and begin to speak after an hour, continuing during five or six hours following. The mushroom addresses the one who eats it, speaking to him of life and death, and the future as well. In addition, it finds lost objects.

The mushroom is eaten raw; the dose varies according to the person and the species. They are customarily taken in pairs and if the individual eats too many he becomes sick, dies or becomes dismayed (*desmaya*), just as happens if he took it in a ritually impure state. They eat them inside the house, and if all is well done, the *curandero* has visions and the mushroom speaks for several hours. The mushroom indicates what made the person sick, and is able to say what the witchcraft was, who did it, on what day, as well as the motive, and can well indicate whether it pertains to a fright (*un espanto*) or sickness that can be cured with other medicines. At times they say he who speaks is Jesus Christ, but the mushroom only speaks Mazatec.

The mushroom predicts whether one is going to die or not and the people believe what it says. Also it indicates who ought to receive part of an inheritance, although sometimes the mushroom can be mistaken. The *curandero* can take alcohol very night (*El curandero puede tomar alcohol muy noche*), although some take it during the rite. The *curanderos* are used to seeing the sea when taking the mushrooms, and one is able to take them only with his consent.

The mushroom sprouts with the rains. It should be collected early in the morning. There is no ritual act in collecting it, but it ought to be eaten within twenty-four hours, from mid-day, and taking no alcohol before eating it. It is not sold in the market, and dried it can last from three to six months.

There are distinct ways of performing the rites among *curanderos*. Some shout or sing, while others work silently and barely raising their voice. The materials for the rite, which are *copal*, chocolate, maize, *pisiete* (tobacco), eggs of hen and turkey, macaw feathers, *amate* (bark paper) and several candles, are placed upon the ground in front of the altar. A large part of the ceremony consists of the manipulation of these materials. To take the mushroom, the *curandero* kneels before the altar, chewing the raw, fresh mushrooms in pairs, one and the other of each pair, until finishing the fourteen.[38] Some questions are presented for which he has petitioned its aid, and he soon sees what it can say and predict.

Fortunately we have excellent data from Huautla which can be compared with each other,[39] theirs being another ceremony distinct from the one which we have been citing. More people participate, although usually they are of the same family. The manner of participation is a most distinctive factor, because the mushroom is shared among all, in a manner that all participate in the experience. The *curandera* represents the others, but they form an integral part of the ceremony by experiencing it directly.

The ritual articles also vary, putting upon the altar images of two saints. In addition, they put flowers, a crucifix, candles and a votive candle, *copal* in an incense burner, and some cups.

The healer apportions thirteen pairs of mushrooms, but they eat them one at a time not in pairs. Before beginning the rite everyone drinks chocolate.

The *curandera* speaks and sings for many hours, during which time she receives help from her daughter. Later the *curandera* performs a dance that lasts two hours. During the ceremony she drinks *aguardiente*[40] and at times invokes the names of Christ, St. Peter, St. Paul or the Holy Spirit. From time to time, a kind of intermission takes place in which the *curandera* and some of the others converse exchanging impressions. The following day one can drink coffee with bread, as long as the fast does not seem to be very strong (*no pareciendo ser muy fuerte el ayuno*).

Among the Zapotec of the coast, you find some distinct details. From time to time the *curanderos* of the region get together, those of greatest age, to discuss important matters; if there is a problem to resolve, one of these men takes the mushroom to obtain counsel.

It is the *curandero* or "wise man" (*"sabio"*) who eats the mushroom for divination. They are prohibited from having sexual relations or drinking alcoholic beverages for four days before ingesting the mushroom, but one can eat and smoke. There are certain indications of specialization in that one of the four species is utilized for the hunt.[41] Each species has its own name.[42]

During the four days prior to the ingestion of the mushroom, the *curandero* keeps it on the altar with the images of the saints. Each day he offers some prayers, after which he washes his face, hands, and feet. Upon gathering the mushroom he makes the sign of the cross, kisses the mushroom and says some prayers. Before eating the mushroom, he returns to the place where it was collected to make an offering, petitioning the deities for more mushrooms the following year. He leaves candles and flowers at the spot.

Among the Mixe the mushroom is referred to in pairs, with a word that designates a sexual pair. The *curandero* does not take the mushroom if he himself is not the one consulting it. He who wants to obtain information consults the mushroom in Mixe or in Spanish. The Mixe word to designate the hallucinogenic mushroom is the morpheme *"hongo"* ("mushroom") to which descriptive elements are added. But there is also a term expressing "that which is borne of itself" (as in Huautla and the Mixteca). The mushrooms are swallowed without chewing, leaving the stalks next to the cross as an offering, and invoking St. John. Only two individuals are usually present, of whom one eats the mushroom and speaks with it while the other keeps watch but does not speak. The dosage varies according to species.[43] Six pairs of one species is prescribed as a children's dose.

Any person can gather the mushrooms, and before taking them they are placed upon the altar of the church while the person lights some candles and *copal* to request the blessing of God and permission to consult the mushroom. They then take them to the house to eat them.

They observe sexual and dietary restrictions that last four days before and after eating the mushrooms, being obligated not to take greasy foods and intoxicating drinks. If these rules are not observed, one can become crazy. A pregnant woman should never eat the mushrooms or else she will become barren (*o también enloquece*); in case of pregnancy, someone else must eat them for her.

In the Chinantla, the *curandera* dries the mushroom and reduces it to crumbs before taking it. When she recovers the forces she begins to divine (*Cuando recupera las fuerzas empieza a adivinar*).

Also in the Valley of Mexico the prescribed manner of taking "the children" (of the water) is by drying them, and they are eaten by "the heavenly worker" (*"el trabajador del cielo"*), as those who know about the mushrooms are called. The patient is not able to communicate with the mushroom as well as the other person. Here the mushroom speaks only in Nahuatl.

In the State of Mexico, the mushrooms are known as *"mujercitas"* (little women), *"hombrecitos"* (little men), *"niñas"* ("girls") or *"niños"* ("boys"). They must be taken mixed but not in pairs. They say that "the girl is stronger than the boy" (*"la niña es más fuerte que el niño"*—Ed.: the meaning here is curious despite uncomplicated grammar; this may be a veiled comparison of or play upon native and imported tradition, since *"el niño"* is also used as a reference to the Christ child). He who wishes to divine takes them without the aid of another person; they are eaten dry and slightly toasted, or well powdered with alcohol or *pulque*. They can be ground on a *metate* or with a grinding stone (*mano*). Also one ought to rub the body with the liquid wherever there is pain. On children, they are rubbed in the same manner, but they are not allowed to eat the mushroom. They are taken at night the better to be able to sleep, the woman after giving birth, and for other questions such as finding lost things, finding the cause of an illness, and to receive advice.

The Mixtec

When comparing the data on the Mixtec with all the others, one can notice that there is a great similarity among them all. No doubt, some points seem to distinguish the Mixtec and they refer to some details of the collection, preparation, ingestion, and properties of the mushroom.

In the first place, all the Mixtec informants affirm that the mushroom has curative power.[44] The mushroom is a very important element in all the towns for which we have data, but rarely do they consider it as a curative. It serves as an agent to foretell the future, speak of the past, and indicate if a sick person is going to live or die; it predicts one's life and the circumstances of death; one can discover the causes of a sickness and prescribe the appropriate cure. The Mixe have a history of attributing a strong curative power to the mushroom. The species from the State of Mexico seems to have strong tranquilizing and therapeutic properties when used to rub on the body and treat wounds. Finally, among the Mixtec there is sufficient agreement about the curative quality of the mushrooms, that one can say this concept characterizes the Mixteca.[45]

There are two points that characterize the preparation of the mushroom, and another characterizing the way they take it, which consequently distinguish the Mixteca towns from others. One of these points is the method of preparing the mushroom. The Mixtec pattern consists of grinding it on the *metate* with water before taking it.[46] In other towns they are accustomed the eat the mushroom whole or in pieces, but not ground up. When one adds liquid as they do in the State of Mexico, they use alcohol or *pulque*. In the State of Mexico they say that they grind it, but this method of preparation is not the only one. However in the Mixteca it constitutes the prescribed method. There the act of grinding it becomes a basic element of the Mixtec rite, and furthermore, it is a unique detail in the sense that it represents a part of the ritual structure that has no counterpart in any of the other regions. The meaning of the method of preparation heightens the importance of taking into account the type of person who grinds it, because it must necessarily be a girl, which also constitutes a necessary element of the Mixtec pattern and a unique part of this pattern. Undoubtedly we are confronted with a belief about the special quality of the hallucinogenic mushroom which is expressed socially through the medium of the type of person who represents complete purity: the girl. We have already referred to the reverent attitude that is displayed toward the mushroom and which has its only parallel at the other end of the social scale in the respect shown towards the elderly.

The gathering of the mushroom demonstrates another unique element; it must be done by a girl. The ritual activities of the Zapotec, which they carry out before taking the mushroom, are represented neither in the Mixteca nor in any other region, but the degree of ritual intensity prior to ingesting the mushroom is probably higher in those two cases. In any case, the necessity for the girl to be present seems to characterize the Mixtec complex and not the others. It again indicates that the mushroom is of a very special quality.

The points that have been detailed which are characteristic of the Mixteca and not other places[47] refer to the method of gathering, preparing and taking the mushroom, as well as to its curative property.

As they constitute a great part of the totality of the ritual, it is interesting to note that all have a feminine component as a point of reference, which can be expressed in the following manner: the mushroom that is taken to cure and divine represents the ritual product of the activities, the tools and the hands of a woman.

General Comparison

When one extends the comparison it is noted that the Mixtec pattern is but one among several distinct complexes. The Zapotec is another, according to which one employs elaborate ritual in the house several days before the ceremony, or the ritual that relates the mushroom with the hunt and deals with the four directions, the gods and lightning; or that by which they request mushrooms for the following year.

In the Mixe pattern there exists a detail by which they bring the mushroom to the altar of the church in order to ask for divine blessing and permission to consult the mushroom. The Huautec (Mazatec) pattern is set apart by some ritual material elements that they employ, by the performance by the *curanderos* and by a distinctive ritual intention. Comparing the special characteristics of the regions,[48] the following observations can be made: great elaboration of rituals before taking the mushroom is what distinguishes the patterns of the Mixtec, the coastal Zapotec, and the Mixe. The distinguishing characteristics of the Huautec include the intention of the ritual, that is, uniting the group by a religious experience, plus the performance by the *curandera* to help bring it about, and the use of certain ritual objects.

Analysis

For the moment it is not possible to explain the significance of these differences nor to discover their causes. In order to know what they signify it is necessary to examine them in relation to several cultural and historical problems. For example: Can we say with certainty that some characteristics distinguish one region and not another? How can we explain them? Do they represent a process of diffusion or cultural loss? Are they due to changes and acculturation occurring as an effect of the Spanish conquest? Or are they related to pre-Hispanic chronology? Certain information needed in order to deal with these and other problems is lacking. We need data on other regions where they continue to use hallucinogenic mushrooms. We must have more extensive accounts from places where there are reports of the mushroom. Knowledge must be extended from the same regions for which we have it to other towns in order to enable us to determine the various forms and establish the linguistic, geographic and cultural connections that indicate where these distinct patterns meet. In order to comprehend the depths of the mushroom complex it is necessary to see it from its cultural foundation, on account of which it is urgent to obtain materials that illustrate the principal points where the practices and beliefs about the mushrooms articulate with other cultural activities. It would be invaluable to have data about other methods of curing and divination within the same community in order to understand how they differ from those of the mushroom. The same can be applied to the historical perspective, for although we have much information from several sources[49] it is certain that we lack an indication of the role the mushroom played in everyday life. That is was important, there can be no doubt.

In all ways one can take advantage of the data and comparisons that have been made. We will make a more detailed analysis that can reveal other characteristics of the pattern and possibly clear up some of the problems.

In the Huautecan usage we note an extension so ample that it raises the question of whether it really is a single form.[50] The paramount character of this doubt is the combination of ritual objects that were used[51] in the ceremony of the *curandero*. They are objects known from some places of native Mexico where they continue to serve as ritual aids from times long ago, usually for divination, without customarily seeing mushrooms among them. They do not constitute part of the ritual objects used in other ceremonies in Huautla, nor do they appear among the ritual regalia in any other case. Nonetheless, they should be considered a unique case and likewise in Huautla since the *curandera* who conducted the different ceremony also knew the use of these other elements. It seems that in Huautla the mushroom has been added to the set of ritual objects pertaining to a form of divination very ancient and widespread in Mexico.

Also attention is called to another form[52] which in some ways seems like a grand ceremony, but which could be considered as distinct in the basic intention, the actions of the *curandera* and the form of participation of those who are present.

The grand ceremony, in which various persons consume the mushroom while the *curandera* does likewise; the dance and the songs of this ceremony are details unique to Huautla. In the complex there appears sufficient diversity in the town, but one does not consider the form using maize and other objects as a part of the formal pattern of the mushroom.

In the Mixteca also it seems there are variations[53] in the use of the mushrooms. Some of the diverse elements of the Mixteca are encountered in other regions. For example, the convention whereby the individual being alone can take the mushroom has also been noted in the State of Mexico, and the one whereby any person can gather the mushroom also occurs in various locations. We should note the Mixteca is similar in some details to other associated regions, but to none when all the details are considered. One views these examples of diversity between the Mixteca and Huautla also as a suggestion of what can be encountered in relation to other regions.

One element that seems to have a value for classification refers to "the one who takes the mushroom." In the Mixteca, among the Mixe and in the State of Mexico, the person who wants to consult the mushroom ingests it. The pattern for Huautla, the Zapotecs and the Chinantla dictates that the *curandero* ingests it on behalf of the other, in order to serve whoever wants the information. In the Valley of Mexico, the "heavenly worker" (*el "trabajador del cielo"*) or the supplicant eats it, but in the latter case the mushroom does not speak. The form in which both eat the mushroom is known only in Huautla.

In the Mixteca and among the Mixe there is somebody who accompanies whoever is going to consult the mushroom, which can be anyone. What is necessary is that their presence or knowledge must provide a reference if it will be the first time the other one is going to take it. The companion need not have command of ritual knowledge.

Two patterns thus seem to emerge, using the criterion of "who takes it." One would be the "supplicant/companion" and the other would be the "*curandero*/supplicant," but it is necessary to consider in the second form that the *curandero* also is a supplicant. The first pattern applies to the Mixtec and the Mixe, and the second to Huautla, Zapotec customs, and the Chinantla.[54] The three usages in Huautla seem to fit the second pattern. Then it will remain to classify the two cases in which the individual takes the mushroom with the presence of neither a companion nor a *curandero*, which corresponds to the State of Mexico and the Mixteca Alta. As they seem quite secular it is necessary to relate them to the first and not the second pattern.

Another criterion which might be mentioned is that of the "moment to emphasize in the ritual of eating the mushroom." Those of the category "supplicant/companion" are characterized by the occurrence of a ceremonial emphasis before taking the mushroom, and afterwards in the other category except in the pattern of the Zapotec which corresponds to the first group in this detail, on account of which it must be classified separately.

Although the Mixtec and Mixe are accustomed to eating the mushroom, note that the companion can be a *curandero* as among the Zapotec or Mazatec. What is important in the Mixtec case is that the *curandero* must unavoidably be present and act, if the supplicant is very ill, although it is always the supplicant who eats the mushroom.

In terms of the presence of the *curandero* as a requirement in the Zapotec, Mazatec and Mixtec cases, the latter comes close to the previous two, that is, the two patterns are alike. Furthermore, considering "the moment to emphasize in the ritual of eating the mushroom," the Mixtec and Zapotec types are more closely related, but they differ in how much they contain of these previous rituals. Comparing the ritual objects that are employed in the ceremonies, those of the Mixtec and the Mazatec appear more similar. Employing the cited character-istics to compare the two patterns, one notes that the Zapotec and the Mazatec do not depart from the *curandero*/supplicant pattern, whereas the Mixtec seem more flexible in regard to who can serve as the companion.

The details of the Huautec have been classified as corresponding to the *curandero*/supplicant pattern, that is, where the *curandero* eats the mushrooms in the presence of the supplicants. It now remains to consider the Huautec cere-mony in which the supplicants as well as the *curandero* take the mushroom. The divinatory function does not suffer, since each directs a question toward the mushroom. Nevertheless it is a special complex by the fact that the assistants, at the core, literally accompany the *curandero*. Furthermore, he conducts it in a special way with dance, some songs and several distinctive words, so he serves as leader and is separate from the others, although all consume the mushroom. To classify this ceremony with the others would be to seriously ignore its re-markable characteristics and what would seem to be its meaning. Such meaning could be defined as the providing of an opportunity to share in an experience wherein each person experiences it, but as a part of the group. The group be-comes the main focus within which each one heightens his experience by being able to share it with the others assisting. The remainder of the ceremonies is characterized by an orientation toward divination mediated by the participation of an individual, always being able to attend others, but as witnesses. From this analysis, the need to utilize a category of "meaning" arises to better understand and more accurately classify the diverse complexes of the mushroom.

IV. Conclusion

The purpose of the essay has been to present a comparative view of the be-liefs, the uses and the rituals concerning the hallucinogenic mushroom in present day Mexico. Some points about this can be briefly summarized.

Within each region from which we have comparable data there appear vari-ants, some small, others larger. Here it is one detail, there another that varies. The same thing occurs between two or more regions, so that each region appears somewhat distinctive, but at the same time resembles in some characteristics a neighboring region or some other, more distant one. It happens more often that a variant of a region is more like that of a far distant place, than it is to another variant near the same region. Among the regions for which we have notes it is doubtful that any have a homogeneous pattern. Huautla serves as a good exam-ple of the differences that can be encountered in a single town.

The known variation does not indicate that the distinctions are infinite. It is probable that we are already on the verge of distinguishing their limits, with places remaining where you can make one or another new divergence fit.

Two fundamental patterns have been formulated based on meaning; one consists of the experience of the supernatural ceremony, and the other is prag-matic, focused toward divination. Using several criteria to analyze the distinc-tions more thoroughly we may distinguish yet others, but in fact they need not all be considered as separate patterns, but rather as variants of a major pattern, of which there are perhaps between three and five.

It is probable that these multiple variants correspond to a central pattern, but it is doubtful it can be found at present, and it is even possible it did not exist in the time of the conquest, in view of the well known regional differences relating to other cultural aspects. By that era people could have found various ways to express themselves as appear nowadays. It could be that a major ceremony such as that of Huautla was celebrated in public and openly before the Spanish conquest.[55] Although the possibility exists that a ceremony of such importance extended over a wider geographic range than the one it occupies nowadays, there is no certainty that it was the only manifestation of the use of the mushroom. It would be more in accord with what is known about pre-Hispanic cultures and culture in general, to suggest that no cultural region is in itself characterized by a single pattern, as already noted. Nevertheless, we know more about the religious life and the social group than the daily life of the people, and less still about the lives of those isolated from the ritual and commercial centers, in whose case it is no great venture to suppose that some of their customs, among those already ancient, have been conserved over time, as they occur nowadays. Nobody knows the pre-Hispanic extension of the priestly hierarchy as far as its control over remote regions. It is very possible that the *curanderos* have existed in all the towns and have executed their office using ritual objects not all that different from those of the priests. And it is also possible that not everything has been lost during the last centuries. Mainly, one has to allow for the possible existence of two or more different patterns in the same place, among the cultures considered, as we have done here based on the data from the regions which have been treated in this study.

Another conclusion that follows from this analysis refers to the character of the mushroom; its use for divination allows it to be classified among other methods for divination, but one must not think that they are equals. The hallucinogenic power of the mushroom basically distinguishes it from other methods. One cannot deny that the beliefs are very deep and the power of faith equal in all methods of divination, but the experience that the mushroom provides has no parallel among other recourses.[56]

Review of the archeological sources on the mushroom has not been attempted, but it is necessary to make note of an interesting detail. By means of the investigations carried out this year in the Mixteca, in the present study a complex has been inferred that seems to be distinctive, that is, has so far not been observed in other places where knowledge of the mushroom exists. This complex is characterized by the fact that it is a girl who must gather the mushroom, and later ritually grind it with water on the *metate*, before the supplicant takes it.

For a long time we have known of the "mushroom stones" of Guatemala, Chiapas and El Salvador.[57] Ultimately found in the hands of a private collector were nine miniature *metates* with nine associated "mushroom stones" which came from Kaminaljuyu and seem to correspond to the pre-Classic horizon.[58] The interpretation that has been made is that it was an offering.

What can we infer as a respectable age for the pattern that now characterizes part of the Mixteca, but which according to these data, could have reached a much greater extension? At the same time it would be a remarkable case of persistence of a ritual element over so many centuries. It remains to be known also if the mushroom was collected by a girl and to know the other details of the facts to call it complex.

Greater efforts are needed to formulate and clarify the results of this test. Other, more meticulous studies should be carried out on the distribution and uses of the mushroom, from a geographical as well as a cultural point of view. An analysis of the present data is needed to relate them to the historical and archeological sources. There is a significant lack of study providing extensive data on the content of the hallucinations of those who take the mushroom, later to apply them to a psychological study of its cultural relations in the villages (*pueblos*), knowledge that can be of great value as far as the problems of mental and public health, and not only in Mexico.

Notes

1. For the most part, the works reflect the great interest and persistence of R. Gordon Wasson and his wife. Their faithful efforts have produced several articles and two books of great merit (Wasson and Wasson 1957; Heim and Wasson 1958). In them you can find the data about the people and circumstances responsible for the recent "discovery" of the hallucinogenic mushroom, including in addition the principal historic sources. The ethnographic details that they provide about the mushroom, such as the beliefs, the uses, the ceremonies and other additional factors, constitute the basic materials to do a comparative study with the data on the Mixtec that are presented here.

2. For botanical sources, consult Heim and Wasson, 1958; in Wasson and Wasson you will find data on the distribution, beliefs and attitudes concerning mushrooms worldwide, in addition to some interpretations. As a noteworthy summary of works by another writer, you should consult the study by La Barre, 1960.

3. This is the term that we will use, without making distinctions between the various species. It is to be noted that the species used in the Mixtec ceremony was *Psilocybe mexicana* Heim.

4. See Fig. 1 (Ed.: Fig. 1 shows a map of the Mixteca).

5. The data on the Mixtec cultures were gathered in the years 1955, 1956, 1957, and 1960, that is to say, in a period totaling a year and a half. One began to gather data on the mushroom in the 1956 season. The work presented here would not have been possible without the assistance of R. G. Wasson, with whom the author made the expedition undertaken June 22 to July 6, 1960. The work that best summarizes the historical and geographical data on the Mixteca is Dahlgren, 1954.

6. There is one reference to the Mixteca Alta (Heim and Wasson, 1958, p. 40), but until now there has been no other report.

7. There could be commerce in mushrooms between the two regions and the Coast, but to date this is not known with precision.

8. Mixteco is classified in the Macro-Otomangue group.

9. Tamayo (1949) speaks of the relationship between altitude and ecological zones.

10. Schultes (1941) affirms that almost the entire variety of vegetation of tropical Mesoamerica is found in Oaxaca.

11. For example, the sacrifice of an animal to thank the Earth for the harvest, or to ritually petition in caves for rain before planting.

12. In other parts of the Mixteca they use *ololiuhqui* or other substances.

13. The mushroom speaks mainly in Mixteco, but also knows Spanish.

14. Also it is said to be "dangerous" or "delicate;" when the mushroom begins to speak one ought not look toward the place the voice is coming from, because otherwise the individual becomes insane.

15. It is interesting to note the parallel of the social forms of antiquity with the *compadre* and *mayordomo*, whose sanctions also have a supernatural base and whose forces, although each is different, control special knowledge. It is remarkable, therefore, that the position which they occupy in society is the highest.

16. If they cite differences, they are few; one says at nine o'clock, another says at ten, and another says midnight.

17. That is, so it activates the senses more.

18. Or like the village loudspeaker that announces some movie or plays "*Las Mañanitas*" (Ed.—A 'happy birthday' song, customarily sung by family and well-wishers in the early morning outside the bedroom window of the celebrant, with lyrics to the effect of "get up sleepy head, the sun has already risen..."). The use of the mushroom is not restricted to the most isolated towns. It is certain that noises momentarily interrupt it, but one gets used to returning to reality from time to time, even without noise being made.

19. "It's like dreaming, or being drunk."

20. In the cases known, having a single companion is commonest.

21. But it must be a person who knows what the correct amount is. Only a few people know this, and not all the *curanderos* know what to do with the mushroom. Whoever has taken it can indicate the proceedings, but not all are inclined towards it, some by lack of interest or fear of this "delicate" mushroom, that is to say, of what a dangerous thing it is.

22. It may be than an important variant is apparent here: there is a place in the Mixteca Alta where the person can take the mushroom even if they are alone, taking it at dawn. He crushes the mushroom with any clean stone. The reasons for taking it are the same, but other factors vary which appear to from a pattern widely distributed in the Mixteca.

23. The shepherds call them *"jongo"*; in Spanish the Mixtecos designate them as "what cures" or "those that are born alone." In Mixtec it is called *si?i*, or *shii*.

24. In this way a locality becomes the center of distribution of the mushroom for the region, and in a certain way can express the tendency to become a ritual center on account of this element.

25. The lack of rains explained the shortage of the mushroom in the 1960 season, although it rained adequately for the maize harvest. We obtained a small amount of the fungi, just enough so that one ceremony could be conducted.

26. The greater the abstinence, the more effective the mushroom is, according to those who comment.

27. When the *curandero* is present, he directs the others.

28. This scheme illustrates the arrangement on the occasion when R. G. Wasson and the author of this study participated. According to other descriptions provided, this form is not outside of the common, but the details that appear here are of the ritual at which he was present

29. On this occasion it was a *curandera*.

30. The mushrooms are counted in pairs, each made up of a male and a female. The dose is commonly seven pairs, but it is increased to fourteen when the brain is "stronger." But "if one goes to take it, he is dazed very ugly or can remain crazy" (*"se ataranta muy feo o puede quedarse loco"*). It is not taken with brandy, but the potency of the dose must be equal to the amount of brandy one can hold. Three pairs are prescribed for children younger than thirteen. A shepherd said that the fungus speaks in Mexican and in Spanish. One must taken them in increments of seven pairs, that is to say, seven are taken; if more is desired, it must be fourteen or twenty-one.

31. Some say that it must be "holy water."

32. Only one informant said that a better result is obtained by chewing the mushroom.

33. To clear away all impurities. If one does not do this, the one who takes the mushroom can become ill.

34. It is evident that they treat the mushroom in a special way. In the case of the *curandera*, this is done in a strong voice and she is apt to speak in a more vigorous manner than is customary among the women, but throughout the time the ritual lasts, the voice of the *curandera* can hardly be heard, one cannot distinguish the majority of the words, except when the voice is raised a little bit to direct some words to others present. There are two classes of mushrooms used for divination: one from the plains, and the one of the manure piles, which can be of cow, horse or donkey. The one of the plains is better because the other is very strong, since "it is very delicate to eat it, people become crazy," and, "one can see the animal thrown over one" (*"uno ve al animal que se echa sobre uno"*). As to its effectiveness, R. G. Wasson indicates that indeed, in agreement with Dr. Heim, one can distinguish them chemically. People say that they can increase the effectiveness by taking the mushroom seven times over two months, and also it is good if the ingestion coincides with the *fiesta* days.

35. The effect takes place in a half hour more or less and lasts three to five hours. The *curandera* uses the alcohol when the supplicant sweats too much, or when he takes more than five hours to return to himself, in which case she blows on the individual and rubs his body.

36. In Heim and Wasson, 1958, Chap. II, the data are presented in detail.

37. Ibid. The data from the Mazatec region are used primarily because they are the best known. Those of other regions are used when possible.

38. *Psilocybe mexicana*, the same species among the Mixtec.

39. Heim and Wasson, 1958, pp. 55-75.

40. Ibid. Wasson (p. 73) writes that the ceremonies of the Huautec were different. On a different occasion the same *curandera* changed the procedure, in that there was no dance and she spoke in another manner. It seems that more was concentrated on the treatment, and the ceremonial procedure was simplified.

41. The candles, the points of the compass and several deities associated with lightning and the four cardinal points of the compass.

42. *Rivea corymbosa* is used when the mushroom is scarce.

43. They say that each species has its own power, that is, to cure, to divine, or to drive crazy.

44. The mushroom says to one: "Take me and I will alleviate you." There are several cases that illustrate its curative power.

45. It is better to consider it as a phenomenon of emphasis or degree, not unique.

46. See footnote 32.

47. That is to say, those that are separate from the general pattern.

48. We compare only these four because they are the ones for which we have the most complete data.

49. In Heim and Wasson, 1958, pp. 17-34, the texts in Nahuatl are included.

50. Wasson quite rightly indicates that there are two distinct methods of divination (Heim and Wasson, p. 73)

51. Refer to Fig. 2, the ritual objects and their arrangement on the altar for the Mixtec mushroom ceremony.

52. See footnote 40.

53. See footnote 22.

54. Although due to serious data being scarce it would be better not to classify the last one.

55. As suggested very reasonably by Wasson (Wasson and Heim, 1958, see dates 6, 7A, p. 38), perhaps it does not suit the most current tastes.

56. Apart from other hallucinogens.

57. Wasson and Wasson, 1957, pp. 275-285; Heim and Wasson, 1958, pp. 113-121.

58. The data are in a manuscript of January, 1960 by S. F. de Borhegyi. I thank Mr. Wasson very much for having sent me these data.

References

Borhegyi, S. F. 1960. Ms. Miniature mushroom stones from Guatemala.

Dahlgren, B. 1954. *La Mixteca*. México.

Heim, R., and R. G. Wasson. 1958. *Les Champignons hallucinogènes du Mexique*. Paris.

La Barre, W. 1960. Twenty years of peyote studies. *Current Anthropology* 1.

Ravicz, R. 1958. Ms. A comparative study of selected aspects of Mixtec social organization.

Schultes, R. E. 1941. Ms. Economic aspects of the flora of northeastern Oaxaca, Mexico.

Tamayo, J. L. 1949. *Geografía General de México*. México.

Wasson, V. P., and R. G. Wasson. 1957. *Mushrooms, Russia and History*. New York.

Chapter Five
The Mixe Tonalamatl *and the Sacred Mushrooms* by Walter S. Miller

Walter S. Miller was a minister affiliated in the 1950's with the Summer Institute of Linguistics, centered in Mitla, Oaxaca. He had lived among the Mixe Indians and was expert in their language and customs, as noted by Allan Richardson (1990, 194), the photographer whose unforgettable images of the Mazatec mushrooms ceremony graced the May 13, 1957 issue of *LIFE*. This was an exceedingly rare expertise and it led to Miller's involvement with Wasson's research in Mexico.

In January, 1954, Robert Weitlaner wrote Wasson to report that his daughter, Irmgard Weitlaner Johnson, had just returned from Mixe territory. There, she had collected information about two types of sacred mushroom from a *curandero* in the village of San Juan Mazatlán. Spurred by this lead, Weitlaner invited Wasson to join forces for an excursion into the Mixe country (or Mixeria as it is known) to conduct follow-up research. So began Wasson's second season of field study in Oaxaca. In regard to the spelling 'Mixe,' Wasson notes that he actually prefers the less favored form, Mije, "because the value of the 'j' is that of standard Spanish, whereas 'x' in Mexico represents any of three consonants" (Wasson and Wasson 1957, 266, ftn).

Shortly after Wasson arrived in Mexico for this work, and before his rendezvous with Weitlaner, he contacted Miller by radio. Richardson, also in attendance for the 1954 expedition, recalls that Miller was initially reluctant to accompany their party because he had only just returned from a leave of absence. Fortunately for ethnomycology, arrangements were made, and Miller's assistance and familiarity with the Mixe proved invaluable. Two of Richardson's evocative photographs from this expedition, showing Wasson, Miller, and Weitlaner, are featured in Reidlinger (1990, 196; and Plate 1, upper photograph).

Richardson recalled Miller from the 1954 expedition as a seemingly remarkable fellow. After observing that Miller was "just as pleasant as expected," Richardson referred to the team's second night, which they spent in the forest, apparently bivouacked. He noted:

> The sky clouded over and looked very threatening. I thought it was going to rain. So I complained to Walter Miller, 'What are we going to do? We have no protection.' And he said, 'Well, I'll pray about it. It'll be all right.' Then he went to the edge of a field and knelt down and prayed. Fifteen minutes later the sky cleared, the stars came out, and everything was fine (1990, 195).

Later in their travels, a local climate of political unrest led Richardson to feel endangered:

> I said to Walter Miller, 'If we're alive in the morning let's get out of here pronto.' Again he replied, 'Well, I'll pray about it. It'll be all right.' So he did. I woke up about 1:00 in the morning in my sleeping bag, and found that nothing had happened: I was still alive (1990, 196).

Miller seems to have possessed other talents that came in handy, judging by the impressions of some. R. G. and V. P. Wasson noted that in San Juan Mazatlán, "sometimes called Mazatlán de los Mijes" (1957, 267), after disappointing efforts to interview Irmgard Weitlaner's original informant, a different *curandero* approached them "with various ailments for treatment: Walter was skillful with massage and manipulation, and also relieved his pains with analgesics, whereupon the old man answered patiently all our inquiries" (1957, 268).

As the text presented below reflects, Miller's field studies yielded interesting results. Even before Wasson contacted him, Miller had already obtained notes from informants in the village of San Lucas Camotlán where he lived, "twenty walking hours west of Mazatlán" (Wasson and Wasson 1957, 273), on several types of mushrooms purported to heal, to work as a form of divination, or drive one crazy. Of one type, Miller noted a belief that eating it induces a dream or vision of two *duendes* or spirits, one male and the other female, who can answer questions about what the future holds, or where to find lost items, and other such matters. Wasson, who first learned of these details from Miller during the 1954 expedition, saw them as an expression of a male/female coupling motif related with the sacred mushrooms, similar to what he had noted among the Mazatec (Wasson and Wasson 1957, 273-274). Miller's preliminary observations from Camotlán, first cited in 1957 by the Wassons, reappear here in the context of a significantly expanded discussion further revealing the depth and complexity of traditions surrounding the sacred mushroom in Mesoamerica.

Through Miller, Wasson soon made contact with Searle Hoogshagen, another field collaborator who provided invaluable data and personal contacts. Hoogshagen, a colleague of Miller's at the Summer Institute of Linguistics, was working in a village of about five hundred residents called Santa María Nativitas Coatlán, ten hours away from Mazatlán by foot (Mazatlán de los Mijes must not be confused with the Mazatlán on Mexico's Pacific coast, a popular tourist beach resort). Shortly after the Wasson team's arrival in Mazatlán, Hoogshagen came calling in person along with two young men from Coatlán, "Severiano Sánchez, age 33, and Cándido Faustino, age 25," both of whom proved "first rate" as informants (Wasson and Wasson 1957, 272). Yet another of Richardson's vivid photographs from the 1954 expedition, taken in Mazatlán, shows these two young Mixe men seated at a small wooden table with Hoogshagen sitting across from them, busily taking interview notes with pencil and paper while Wasson, standing, looks on (Reidlinger 1990, Plate 1, lower photograph). The drama captured in this unassuming 35mm snapshot of historic discovery as it unfolds is illuminated by the realization that these are the individuals cited in the text which follows as CF and SS from Coatlán. Hoogshagen later wrote a significant article titled "Notes on the sacred (narcotic) mushroom from Coatlán, Oaxaca, Mexico" (1959). A species used in Coatlán proved to be new to science, and that same year Roger Heim named it *Psilocybe hoogshagenii*. Hoogshagen had provided the Wasson team with the collections, gathered in July, 1958, which became the basis for the concept and description of the species. The same species was later reported in shamanistic use among the Chinantec, by Rubel and Gettelfinger-Krejci (1976).

Miller continued his collaboration with Wasson after the 1954 expedition, and joined him again in person in 1959 for further field studies in the Mixeria, along with Irmgard Weitlaner Johnson, Wasson's daughter Masha, plus a French contingent including mycologists Roger Heim and Roger Cailleux, and anthropologist Guy Stresser-Péan. The latter recalled that when their entourage arrived in the Mixe village of Zacatepec, there was "a great noisy feast being celebrated by a famous local *cacique*. Mr. Miller was leading us" (1990, 234). Irmgard Weitlaner Johnson has contributed a detailed recollection of this "memorable trip to the Mixeria" (1990, 136).

Below, Miller offers some of his observations on the sacred mushroom among the Mixe, especially in regard to its connection with the Mixe "Book of Days" or *tonalamatl*, a native calendar of ritual significance. The term *tonalamatl* is a compound word, the suffix of which, *amatl*, refers to native bark paper of Mesoamerica, invented independently from the Old World, where the invention of paper is credited to the Chinese circa 100 AD (Simpson and Ogorzaly 2001, 389). The prefix *tonal*, meaning day, is related to *tona* or "fate" (Dow, 1986), reflecting a connection in native thought between one's day of birth and one's destiny, perhaps analogous to the popular horoscopes of Western tradition. Just as the zodiac links one by month to a sort of totem-like calendar figure—the ram, the bull, the twins, etc.—so the day of one's birth in native Mexican tradition links one to an animal or similar personification of nature. The *tona* thus becomes one version of the animal doppleganger motif in Mesoamerica, here taking a benign form. Wasson explains it as:

> the Indian belief, formerly universal, that every human being at the instant of his birth is linked with a particular animal born at the same moment. They live parallel lives and when one is hurt the other suffers a corresponding injury, when one dies the other dies. Various divining methods were used to arrive at

the species of the *tona*, and one was always careful throughout life to avoid injury to animals of the species of one's *tona*. The word *tona* seems to be taken from the Zapotec language, or one of the dialects of Zapotec, but to come ultimately, via Zapotec, from Nahuatl *tonalli*. María Sabina speaks of the *tona* as *soerte*, a corruption of the Spanish *suerte*, 'fate' or 'fortune' ... We see that in María Sabina's world the *tona* still commands full credence (Wasson et al. 1974, xii-xiii).

The Mixe *tonalamatl* presumably has functioned as one of the divinatory methods of which Wasson speaks.

This connection between the sacred mushrooms and the Mixe *tonalamatl* is likely to be new for some readers. Miller notes that among the Mixe, a general term for the sacred mushrooms is *nä:xwin mux* (as Miller spells it in the text presented below), meaning World Mushrooms. The Wassons reported a synonym for this term which they learned, along with Miller, in the village of Mazatlán de los Mijes:

> Manuel Agustín, an old man, disclosed to us a surprising fact ... there is a synonym, *tu:m'uh*. Our friend Felipe was helping Walter ... and Felipe changed the word to *tu:m 'ungk*, which another elder of the town Gerónimo Antonio, later confirmed. They and ... others knew the word and its meaning, and they all agreed that its inner sense was 'that which is born of itself,' *lo que nace por sí mismo*. Here is the same figure that we find in the Mazatec *si³to³*! *Tu:m* of itself means nothing in Mije and occurs elsewhere only in the distinctive Mije calendar of day names, where in the complicated rotation of the native calendar it might turn up in the combination *tu:m 'uh*. The term seems to be an archaism, with mystical implications that will have to be explored by further study (1957, 269-270).

This seems to have been the first glimmer of the connection between the mushrooms and the Mixe *tonalamatl*, and the paper Miller presents here addresses the need for further study cited by the Wassons in the quotation above. Wasson (1980, 235) notes two Mixe accounts previously reported by Miller (1956) referencing hallucinogenic fungi. For further information on the sacred mushroom among the Mixe, see Lipp (1990).

Weitlaner is generally credited as the first field investigator to document the survival of nanacatlsm in Mexico in modern times, based on ethnographic notes he obtained from Huautla de Jiménez, backed up by a voucher collection purporting to be sacred mushrooms, in 1936. Thirty years later, Miller offered the following essay in tribute to him as one in a series of similarly honorary contributions, including one by Wasson on *ololiuhqui* and other hallucinogens of native Mexico. The occasion was Weitlaner's 80th birthday, and the homage or *Homenaje* was edited under the auspices of a committee presided over by the Alfonso Caso, the distinguished Mesoamericanist whose work has also contributed a great deal to understanding of the sacred mushroom complex (Wasson 1973). It is a pleasure to present Miller's tribute here for the first time in English translation.

*　　*　　*

Miller, W. S. 1966. El *tonalamatl* mixe y los hongos sagrados. In *Summa Antropológica en Homenaje a Roberto J. Weitlaner*, ed. Antonio Pompa y Pompa, pp. 317-328. Mexico City: Instituto Nacional de Antropología e Historia, Secretaría de Educacíon Pública.

In honor of my good friend and beloved colleague Mr. Robert J. Weitlaner, it is a pleasure for me to offer some brief notes on two subjects—calendars, and mushrooms, which for many years have been the subject of Weitlaner's perennial research.

Before presenting these notes, I would like to acknowledge that I owe a great deal to Robert. In a world filled with "professional jealousy," spite, and distrust, it has always been very encouraging to find in Robert a desire to help others, and to amicably provide them with all the data, ideas, and advice within his power that would assist them in the rapid progress of their own research. It is not surprising, then, that Robert's colleagues hold him in such high regard. No matter where he has been to study among indigenous peoples, he has been so friendly and nice that many inquire about him with true affection. Robert has set an example for us, one that merits imitation.

Though we are sorry that some of his studies have not been published but remain yet to be published, it is remarkable—it is surprising that he has contributed so much to the fields of linguistics, ethnology, and archaeology. He reminds me of Franz Boas, who was much honored for his multiple and extensive contributions to various branches of anthropology.

Years ago, in San Lucas Camotlán, I acquired the first Mixe *tonalamatl*, the original of which is now in the *Biblioteca de la Nación* ("Algunos Manuscritos y Libros Mixes en el Museo Nacional," *Tlatoani* Vol. 1, No. 2, March/April 1952, pp. 34-35; Cuentos Mixes, I.N.I., Mexico 1956, p. 61—a reproduction of two pages from the *tonalamatl* cited). During that time, I also collected the first data on the mushrooms that we now know as narcotics. Those data did not reveal how the narcotics are used. And even though the data suggested that the mushrooms used were special ones, and were used in order to read that which was secret and were narcotics because they produced "visions," the data do not give any idea as to why the mushrooms were considered sacred. The following is what I have gathered:

The Mixe know various types of mushrooms. There is an orange one they call *nanacate* which they consume as a food. There are others that appear to be narcotics, and which are used for other purposes. JT gives us the information that one class of mushrooms has been used as medicine. His nephew Alfonso was ill for five years, afflicted with an unknown disease. He couldn't walk, but rather scarcely drag himself along. He sat on a little bench inside the house or on the patio. He was given these mushrooms, and was cured and could walk again. JT tells us:

Another type of mushroom puts one to sleep, causing visions. The vision induced is always the same: two dwarfs or elves (*dos enanitos o duendes*), a male
and a female, appear to the one who eats the mushrooms. They speak to him
and answer his questions. They provide him with information as to where lost
things can be found. If he has had anything stolen, these dwarfs or elves identify the thief and the location where the stolen item is hidden. If one plans a
trip, he is told what kind of luck he will have.

Cerilo from Santa Margarita Huitepec has eaten this class of mushrooms on
various occasions. They were not effective the first time. Cerilo has a son named
Delfino. When he (Cerilo) was going to eat the mushrooms, he was afraid that
something would happen to him. For this reason he told Delfino to take care of
him, so that nothing would happen to him. Then he ate them, and the little people (*los chamacos*) indeed appeared to him. He spoke with them, and asked
them about the trip he was going to take, because he had five donkeys and was
going with Delfino to the *Istmo* (Ed.—the easternmost portion of Oaxaca is
known as the *Istmo*). The dwarfs told him not to go because all his donkeys
would die. They told him about many things. Then the male dwarf said: "We
have to leave because the rooster is crowing." (In other words, it was almost
four in the morning.)

The dwarfs disappeared, and he woke up. Then he asked his son if the
rooster had just crowed, and Delfino answered yes. But he (Cerilo) didn't believe the dwarfs and took the trip anyway. Well, truly, just as the dwarfs had told
him, all five of his donkeys died during the trip.

By reflecting now and posing more questions to some of the same informants, I found some of them to be reluctant to disclose the complete data because of the "delicate" nature of the sacred mushrooms and their use.

During the first years of my research on the Mixe, Bernard Bevan told me
about the discoveries of indigenous "calendars" obtained by Robert and his
companions. He advised me to verify whether the Mixes still used such a calendar. Then, he introduced me to Robert, who explained to me a little about the
calendars that had been found so far. Following their advice, I went around always asking for "the times" during which they (the Mixes) sowed, or did other,
different field chores, and insisted that they give me the Mixe names that were
used to indicate "the times." But they always denied remembering such a thing.
Finally, it was my interest in the old books and manuscripts about language,
rather than my role as an inquisitive person, that allowed me to get the *tonalamatl*. But it was years later, in a town far away from there, that I found a relationship or connection between the *tonalamatl* and the sacred mushrooms.

The stories about the mushrooms used in Camotlán resulted in Robert and I
later becoming collaborators on a research excursion to the easternmost part of
the Mixe route. After I told Robert about the mushrooms, he requested my permission to send a copy of the stories to R. Gordon Wasson, with whom he had
begun an extensive study on the use of mushrooms by indigenous Mexicans—in
present times as well as in pre-colonial and colonial times, represented by records (*lienzos*), manuscripts, and old books. In December of 1953, Robert's

daughter Irmgard Weitlaner Johnson was in San Juan Mazatlán conducting her research on everything relating to indigenous woven material. She succeeded in outlining an agricultural calendar and, in addition, discovering that the mushrooms were still being used there. That is why in May of 1954 Robert and I joined Mr. Wasson in going to Mazatlán. By using my photostatic copy of the *tonalamatl* of Camotlán, we managed to outline the *tonalamatl* that was still being used there. Then, using the data Irmgard provided us with, we took another version of the agricultural calendar, and more extensive data pertaining to its significance and uses. We also managed to obtain enough data about the use and effect of the sacred mushrooms, and the crushed seeds of a vine (it appears to be a morning glory species) that are substituted for the mushrooms in times when they are not available.

I do not intend to discuss all of that here. I want to limit myself to the mushrooms and their connection with the *tonalamatl*, presenting additional data that I have been collecting from informants from Zacatepec, Cotzocan, Camotlán, Juquila Mixes (Ed.—not to be confused with the village of Juquila cited in Chapter Seven, *One Step Beyond: The Sacred Mushroom*), Santa Margarita Huitepec, Coatlán and its settlements (or laborers' quarters), and from San José Paraíso. I thank Searle Hoogshagen and Norman Nordell, my colleagues from the Summer Institute of Linguistics, for having provided for my use their unpublished data on the present use of the *tonalamatl* in their towns—Hoogshagen being from Coatlán, and Nordell from San Juan Guichicovi. (Nordell also found data on the agricultural calendar.) In addition to Hoogshagen's data, I managed to collect some through my own research, with informants from Coatlán and from San José Paraíso.

During our research in Mazatlán, our informants were men in their old age, 65 and 67 years old, and who were well respected "*brujos*." The oldest of them all was *mayordomo* of Patron Saints' Day. Though Irmgard had gotten data with another medicine man, and with his brother, they refused to give more data. In the beginning, GA, who came to be our principal informant, also denied knowing about the "calendar." But upon seeing the copy of the *tonalamatl* from Camotlán, and hearing me read the names, he started smiling and said: "yesterday was 'kø'ønt huiky' (11 tabaco), so now we are at 'kø'øx pä:' (12 raised or upright pole)." (Ed.—Assuming GA is the actual monogram, this would appear to be the informant cited by Wasson and Wasson, 1957, p. 269, as Gerónimo Antonio). We got the data about the mushrooms from these two principal informants and from various other *brujos*. It seems to me that these medicine men and medicine women deserve to be called priests and priestesses for lack of better descriptive terms, since they appear to be the guardians or "repositories" of knowledge of the *tonalamatl* and the agricultural calendar. And from what we have seen and experienced (and been shown), I conclude that they are also the guardians of knowledge about the sacred mushrooms.

Even though we inquired and learned a lot in Mazatlán, the data we collected then suggest many more questions that we would like to ask. These questions are the result of thorough study and reflection upon the collected data. We did not come up with these questions all at once, because we were very busy writing down in Spanish the answers that were given in response to our questions, from what I was able to understand from the intermittent conversations in Mixe among the informants themselves. It will be valuable to briefly state what they have told us about the mushrooms.

The narcotic mushrooms are considered sacred.

Those used for divination and healing are not common and abundant, but rather special ones. There are three types, each having its peculiar name. Altogether, these three types are given the special name "nä:xwin mux"— 'mushrooms of the World.' While CF and SS from Coatlán were arguing about how to say things in Spanish, I heard one of them say "nwintsøn'ähtøm näxwin mux"—'our Lords, mushrooms of the World.' The term "our Lords" is the same as the title used in the church to refer to Jesus Christ! There can be no doubt that the designation in Castillian "*hongos sagrados*" captures the sense in which the people revered these mushrooms.

A special name is found in the *tonalamatl.*

GA from Mazatlán did not give us any equivalent names for some of the names in the *tonalamatl*. MA did give us the name equivalent for one of them. This information turned out to be of great importance. The word was "uh"— Earth or World. This same informant, while talking about the names they give to the sacred mushrooms, said that these were also known as "tu:m'uh" or, according to other people, as "tu:m'unk". The informant said, however, that the first form was the correct one. And it is in this word that one finds the connection between the *tonalamatl* and the mushrooms, because "tu:m'uh" (or "tu'um'uh") is one of the days of the *tonalamatl*, which means "World one." Those who use "t:um'unk"—child one—say that it is so named because "he is only born of the Earth." They believe the World sends them. "Suddenly, there he is" they say.

The season of the sacred mushrooms depends upon thunder and lightning.

Regarding the season of the mushrooms, informants were unanimous in pointing to the same season— after the first showers. SS from Coatlán said:

"Nä:xwin mux" is not found in just any weather: the mushrooms appear after a downpour that soaks the earth, accompanied by thunder and lightning. (Subsequently, all the informants from other towns corroborated this.) The season lasts as long as the downpours with thunder and lightning continue, sometimes fifteen or twenty days, sometimes an entire month. Regularly, it is in the month of June—or it starts around the middle or the end of June, extending a little into July.

TQ from Mazatlán said:

Its season is in June and July. There is an abundance then. They are also found in September and October, but by then they are more scarce.

The mushrooms may be used fresh or dried.

Usually, the mushrooms are taken fresh; but TQ from Mazatlán said:

> They may be dried. They are put outside in the sun—for one day on a hot day; two days if there is not much heat. They last no longer than fifteen days in a month.

It is said that it is also possible to dry them near a fire; but at a distance sufficient not to burn them. Because the mushrooms appear in rainy seasons, they run the risk of "putrifying" ("*posca guarse*"). This is because the humidity then is higher than ever. There are days when it reaches around 90 to 95 percent. It is said that the effectiveness of dried mushrooms will last one to two weeks at most. On rare occasions, and with much luck (with great care or with less humidity in the air) they last for almost a month; but it is impossible to keep them for months or for a year, as some foreigners have wanted to do.

They soak the mushrooms only when they have been dried, and in the case of being fresh they do not soak them, but rather use them as they are.

There are rules for whoever wants to use them.

There are rules that govern everything related to the mushrooms; picking them, the preparation of the one who will eat them, his ingestion of them, and recuperation afterwards. That is, the sacred mushrooms are taken in accordance with a rite that lasts seven days, not just one noon and night of preparation, as foreigners have wanted when eating the mushrooms for their experiments. Taking the mushrooms is a very delicate matter, since they are very sacred. It is not a matter of whim or momentary passing fancy as if one were taking an aspirin or Alkaseltzer. TQ from Mazatlán said:

> There is no special preparation, nor is there any fixed hour to look for and collect the mushroom. One can give thanks to God for the mushroom at the moment of collecting it. After collecting it, if one wishes, he may stop by the church—carry the mushrooms in a gourd on the altar, and plead with god and the Saints to give the mushrooms effectiveness. Then he takes them home where he will eat them.

But in order to eat the mushrooms, there is a special preparation. Whoever is going to eat the mushrooms has to be on a four-day diet prior to the day of eating them. During these four days he cannot drink coffee, liquor, and cannot eat chicken (nor any type of bird), eggs, grease, nor pork. However, beef, beans, tortillas, *atole*, maize stew, and cheese may be eaten. Whoever is on this diet may not have any relations with a woman during these four days.

On the day he is going to eat the mushrooms, he may eat a quarter of a tortilla, his *atole*, and a piece of cheese upon awakening at dawn. That is all. Then he eats nothing else during the entire day.

After eating the mushrooms, one has to follow the same diet for four days, just as before.

If one eats them without observing the preparation, he will become deranged; he won't recover any more—he won't be the way he used to be before; he will remain crazy.

The mushrooms are taken at "møh ko:ts"—between 10 and 11 pm—and the effect lasts until "hobø'ø"—dawn, four in the morning. When beginning to eat a mushroom, one is accompanied by one or two friends, with a candle. Then, the mushroom lets whoever ate it know in his mind whether more candle light is needed, or what is needed to cure the sick.

When the mushroom is done "talking," and everything is over, only the one who ate it stays there until the afternoon of another day, since one remains limp and unconscious. He who is lucky eating them always goes to heaven and sees there how he is ordained by God and "nä:xwin"—the World.

MA from Mazatlán said:

> Before eating, at about the hour of prayer, all the mushrooms are taken to the church, and the patron saint is asked for forgiveness and permission to eat the mushrooms. At about seven o'clock in the evening they are eaten. At about three o'clock a.m., one takes a shower. At the break of day, the mushroom stalks, together with a candle, are taken to the cross.

SS from Coatlán said:

> When eaten, the mushrooms are taken in pairs. One is male and the other is female. (They explained that a total dose per person is calculated by pairs—and NOT that they put them in their mouth by pairs, as the Mazatecans do.)

Before eating the mushrooms, they are placed upon a table before the altar (a little lower, however), and incense made of the "po:mkipy" wood is burned. The incense itself is called "po:m" (probably *copal*—Ed.). Then they worship the mushrooms, and plead with the mushrooms to tell them that which they want to know.

Whoever eats the mushrooms is taken care of, so that nothing can happen to him. First, he lies down, waiting for the mushrooms to take effect. Then, once the mushrooms begin to "work," the "eater" can get up to walk and talk. But those who are taking care of him won't touch him or talk to him. If he wants to go out to the patio, no one is to prevent him from doing so. But it is the mushroom itself that demands that he not go far away. He goes out only to get some air and refresh himself, and then he comes back inside to continue chatting.

They always try to eat the mushrooms in an unoccupied house, near the edge of town if possible, so that there will be no noise that would disturb the one who eats the mushrooms.

The body is left weak, because if the body were ill the mushrooms "massage" the illness as it were. ("To massage the illness" might seem to indicate that someone gives the ill one a massage. It is NOT so. This means, the day after having eaten the mushrooms, the person's body feels as if it had gotten a massive massage.) And even if he had been completely healthy when eating the mushrooms, he would still be unconscious (or weak). Therefore, he has to take good care of himself—for some three days, he may not eat meat or chili.

Note: In Juquila Mixes, after massaging a sick one, the massaged person is considered as being in a delicate state, and has to take good care of himself so as not to hurt himself, since this could be permanent. He has to take care not to consume any liquor; he is on a diet, as far as certain "hot," "cold," or "heavy" foods are concerned. He stays in bed, and if he has to leave the house, he covers himself well (up to his head) in a blanket—even when there is sunlight and it is hot.

There are prohibitions and warnings for those who want to eat the mushrooms.

According to SS from Coatlán:

> A pregnant woman may not eat the mushrooms once her pregnancy shows. That is to say, starting from the second or third month, she may not eat them. I don't believe there is any danger before this time, but if they are eaten after the second or third month she will go mad—she won't come out good again.

If a woman is pregnant and has a craving for the mushrooms, she may look for another woman to eat them for her. That is why they plead with the mushrooms to show their effect for the pregnant woman. Her husband could very well eat them for her. (That is to say, it is not necessary that a woman eat the mushrooms for her. Either her husband or some other family member can eat them for her.)

TQ from Mazatlán said:

> If the mushroom eater were disrespectful at any time, and said: "That mushroom, what good is it? It is good for nothing!" –then the mushroom would punish him by giving him visions of snakes, jaguars, etc.

SS from Coatlán said:

> Whoever eats the mushroom becomes like one sleeping, and sees things as if he were dreaming. He sees many animals, but what he sees mainly are snakes.

Note: They assume that there are some snakes and jaguars that are super-natural beings. Therefore, to see them in the visions produced by the mushrooms is considered a punishment, and not a hallucination of nature like one of those seen by those who go too far drinking alcohol or *mezcal*.

If something were to scare whoever ate the mushrooms, and its words got blocked, then the mushroom stalks are again placed in front of the altar. Then they burn incense again for the mushroom stalks and worship them, and ask them for forgiveness (they ask the mushrooms to forgive the affected ones)—and ask the mushrooms to do the one who eats them good. Then, the affected one is given a bowl of chile salsa to drink so that he may turn out well.

They give various reasons for the number of pairs that one eats.

TQ from Mazatlán said:

> Before eating the mushrooms, one has to know how much *mezcal* it takes to get drunk, in order to take the measure of the dose of mushrooms one can eat. For example, I eat six pairs of the "pi:tpa" kind, but other people eat eight, ten, or twelve pairs. I eat three pairs of the "ätkä:t" kind, but others eat five, six pairs. Of the "ko:nk" I eat a whole one or the edge of a pair—no one eats more. If only one of the "ko:nk" kind is found, then the entire cap (*capuchón*) is eaten. If a pair is found, then only the edges of the caps are eaten. But no one eats more, because it is "ko:nk".

Besides this method of calculating the number of mushroom pairs, some-thing very interesting came up in the chat and explanation of SS from Coatlán:

> The major employees of the WORLD ("nä:xwin") are thirteen in total; that's why thirteen pairs are eaten. There are nine minor employees ("tunkmø:tpa"), and that's why nine pairs are used for adults; six pairs are used for children.

It is not understood exactly what the informant wanted to say. It could be something like what the Zapotecans from Mitla tell, and apparently the Zoques too—that there is a supernatural world where they also have their affairs organized like a municipality with its authorities and "employees." (The term "em-ployees" [*empleados*] is common in the Mixe towns from Coatlán to Juquila Mixes. "Major employees" and "minor employees" are common terms in desig-nating the degrees of various offices—Sheriff [*topil*], Mayor [*mayor*], "Judge" [*"juez"*], Alderman [*regidor*], Syndic [*síndico*], Secretary [*secretario*], Justice of the Peace [*alcalde*], and President [*presidente*].)

Notice also that thirteen is the count (*cantidad*) of numbers (*números*) used in the *tonalamatl*—thirteen numbers with twenty names. Could they be related?

Here there is room for more research. Who are these thirteen major employ-ees, and nine minor ones? What are their names? What type of positions do they hold? What relationships do they have with our world or material and social order, with the men and women that live in the Mixes towns? These and many other questions we would like to ask.

What is the origin or source (*proveniencia*) of the use of sacred mushrooms of the Mixe *tonalamatl*?

As far as the use of narcotic mushrooms is concerned, we already have many data on hand on the use by many races, and almost from the beginning of historic times, thanks to the tireless works of R. Gordon Wasson and colleagues and collaborators. However, we do not have any data that allow us to point out what the origin is of the use of these mushrooms among the Mixes. From certain towns, we have reports indicating who sowed the knowledge of the mushrooms in them, making them common property. But that does not give us any idea of the origin of the custom that has been kept by the medicine men "priests" (*"sacerdotes" brujos*) since long ago, in the same towns.

Could it be possible that studying the *tonalamatl* would give us an idea of the origin of the sacred mushrooms? Perhaps yes, if only we knew which dialect the *tonalamatl* was written in. In any case, we would then have to know where the use of the name 'tu:m 'uh" or "tu'um 'uh" as a designation also for the sacred mushrooms came from.

Some of the twenty names are words of almost daily and common use, and also appear in almost all dialects, in slightly varied form—for instance: tooth, reed, jaguar, tobacco, grass, snake. Others such as "uh", "jaimy", "tap" are not used except in the *tonalamatl*. What is more noteworthy is that the numbers are not those used in daily affairs. They constitute a special mode of counting. The forms that are used in the *tonalamatl* for eleven, twelve, and thirteen are used neither as cardinal nor ordinal numbers in any of the dialects for which I have data. Varying with the dialect to which we compare them, some of the remaining ten terms show a relationship with, but are not equal to, the cardinal numbers. For example: in the Camotlán dialect, the numbers 1, 2, 4, 5, 9, and 10 from the *tonalamatl* show a relationship to the ordinary numbers. The correspondences vary according to the dialect. In data provided by John Crawford, pertaining to Tonaguía: for the number 16, "äpädu'um" is used, which appears to be 15 plus 1, where "-du'um" is 1, and is the combining form of "tu'um" from the *tonalamatl*. The number 18 is rendered as "äpädu:k" which appears to be 15 plus 3, where "-du:k" is 3, and is the combining form of "tu:k" from the *tonalamatl*.

We lack comparative data to allow for answering the questions: Does the *tonalamatl* use the common language of a certain Mixe dialect? Or perhaps could it have been a special priestly language, and the numbers a ritual method of counting?

The Mixes' customs and experiences of eating the mushrooms contrast with those of the Mazatecans.

The Mixes as well as the Mazatecans eat narcotic mushrooms. The Mixes consider these mushrooms to be sacred, and apparently so do the Mazatecans. Apart from that, almost everything else is in contrast. Among the Mixes, the mushroom is never eaten in groups of people, as Wasson observed among the Mazatecans. Neither do the Mixes put the mushrooms in their mouth in pairs. The Mixes use incense and candles in their reverence for the mushrooms before eating them. But they do not use accessories for the divination. In this they contrast with the Mazatecans, who use feathers and *cacao*, etc.

It appears to me that we lack descriptions of what the Mazatecans themselves see under the influence of the mushrooms. As I recall, all of the descriptions are of what Wasson and his colleagues saw and felt (or experienced). Therefore, showing any contrast between their descriptions and those of the Mixes is not valid as a basis of contrast between the Mazatecans and the Mixes. However, it won't be in vain to note that the Mixes never talk about seeing geometric figures, nor fantasies of bright and kaleidoscopic colors. Neither do they talk about feeling euphoria such as that which was felt by the investigators mentioned among the Mazatecans.

Wasson reports to us that on one occasion he asked a question about his son. The Mazatecan medicine man, being under the influence of the mushrooms, answered. But Wasson does not tell us how the medicine man arrived at the knowledge. I suppose it wasn't because of seeing geometric figures, or fantasies of colors. It could have been something like what TQ from Mazatlán told us: The mushroom says— ... enters the mind.

Perhaps that is how the Mazatecan medicine man arrived at what he later told Wasson. Therefore, lacking Mazatecan explanations upon which to rely, we should procede slowly to deduce that everything is in contrast between the Mixes and the Mazatecans as far as their experiences with eating the mushrooms are concerned. Nonetheless, there are contrasts that catch the eye. In Wasson's experience concerning his son, he who had eaten the mushrooms continues chatting with the visitors, asking them questions and giving them answers. This is in complete contrast to the Mixes, where the informants—most of whom are medicine men—indicate that those who "accompany" the one who eats the mushrooms must be very silent; and also, that they are NEVER to talk to nor touch him, nor try to impede his movements while he is under the influence of the mushrooms. They also report that he (who is under the influence) pays no mind to the others accompanying him—he never speaks to them. The conversation he has is with the mushrooms.

CF from Coatlán declares:

The mushroom speaks. The person who ate (the mushrooms) poses the question, and the mushroom replies (dialog). But it is all the same voice, and comes from the mouth of the person who ate the mushrooms.

Other informants, though they agree with CF that the one who eats the mushroom is the one who talks, insist that it is NOT all the "same voice" but rather, "the voice changes when the mushroom talks. One notices that it is not the voice of the one who ate the mushrooms." Some insist that sometimes one can hear the female mushroom, and sometimes the male one, when they are answering the questions of the person interested. They also insist that sometimes without asking the mushrooms any questions, they speak and scold one, or give warnings or advice. Those who accompany the one eating the mushrooms also hear the voices. This is truly in contrast to all reports coming from the Mazatecan region.

Informants from Coatlán and Camotlán and their agencies say that some "little people"—a little man and a little woman—like elves, appear to whoever eats the mushrooms (see pp. 2, 3, JT's explanation). It isn't that they are "little people" as far as age is concerned. Neither are they represented as old people. Apparently, they are adults, but small, like "dwarfs"—about 65 to 75 cm in height. We have not heard anything from or among the Mazatecans about this sort of apparition.

Those from Camotlán assure us that, in cases where the eater wanted to guess the whereabouts of some object and everything relating to its loss, the mushrooms made him see "as in a cinema," as if it were happening all over again. In one case, the interested person saw when a neighbor stole the article and took it away to hide it in his own house, in a trunk. The one who ate the mushrooms then immediately left in the direction of that house; and those who were taking care of him had no alternative but to accompany him, in order to prevent anything bad from happening to him. He went directly to where the thief lived, entered the house without saying anything and without greeting, went to the trunk, opened it and took out the stolen item. In the meantime, his companions had to forcibly restrain the owner of the trunk so that he would not attack the one who had eaten the mushrooms. After that, they turned the thief over to the authorities. The same thing also happens in cases of lost items that were not stolen. The mushrooms cause people to see where and how they have lost things, and where those things are right now.

There are times when the mushrooms "take" the person to some place and make him see or do things outside the ordinary. Such was the experience of EG from Coatlán:

> I ate the mushrooms and they took me to this place called "høːmbohtu'u."
> There I saw two big pots covered with a huge stone. I managed to lift the stone off and saw that the pots were filled with silver coins. But there at one side was a snake. I covered the pots again and went home. But when I no longer saw that setting, I did not have the courage to go again and try to get any of that money.

In reality, he had not left his house. This is what he told me when I asked him a few questions. "I did not leave my house. I was ill. That's why I ate the mushrooms." By eating the mushrooms, he had the experience of walking through a certain street in his village and arriving at a place they know by the given name that means "Road of the Fire Wind" (hot wind). They say that there is a small hole there, through which hot wind (air) is always blowing. In some way which he does not explain, "he entered the hill" and that is where he saw the pots. Various stories he wrote down tell about a big snake as the guardian of some treasure. That is not a new theme. What is surprising is that being under the effects of the mushrooms he went to uncover the pots, in spite of the snake's presence. Of all the stories of this type, this is the only one in which the actor has done such a thing. In all the others, he ran away from the place upon seeing the snake.

Some mistaken impressions are corrected.

These so-called hallucinogenic "sacred mushrooms" have captured the imaginations of all kinds of people and many countries. There are many people who have written articles on these mushrooms. One notices that the terms applied to the mushrooms by some of these people are not in agreement with the data and reports given by the Mixes. Neither are the names given to the indigenous rites that are connected with the mushrooms. So it is evident that those terms or names cannot be taken into account universally for all usages of narcotic mushrooms. It is essential that we use our terminology in the correct sense.

I want to make the following observations, for the purpose of correcting some mistaken impressions.

We read and hear people talk about "the mushroom cult in Mexico and Central America" as if there were one unity; as if the use of mushrooms by any race and under any circumstances formed a part of a totality or religious unity going beyond linguistic, cultural, and geographical barriers. The data alone that we have just considered here give an idea of how false this position is. What we can say is that several indigenous races, in their separate ways, do give credit to some of the mushrooms we consider sacred for being intelligent.

Doctors Heim from Paris and Guzmán from Mexico have identified the different narcotic mushrooms picked so far. It is obvious that one is not dealing with the same mushrooms in all cases. The Mazatecans talk about three, or possibly up to four types. The Mixes talk about three types, and it is possible that they include one or more not used by the Mazatecans. And furthermore, it is not certain that the mushrooms used by those from Mazatlán are the same as those used by the people of Camotlán and other villages. And additional data make it probable that some of the mushrooms used in Cotzocon, Zacatepec and Alotepec are of yet another variety. In the same way also, the rites connected with the eating of the mushrooms vary from place to place in the Mixe direction. Besides, we have already seen the considerable contrast between Mixes and Mazatecans, as far as rites are concerned.

With the collected data we can say the following:

There are indigenous people in Mexico and some Central American countries who know and use various types of mushrooms that prove to be narcotics. Some tribes consider them "sacred" and observe prescribed rites for their use, and even venerate them, burn candles and incense to them, and make petitions and prayers to them or to their stalks. In some cases, small groups eat the mushrooms. In other cases, only the "priest" or medicine man eats them—sometimes for himself and sometimes in a vicarious manner for those who come in for consultation. In other cases, the use is "lay," private and practiced by a single person.

Another word that many people use inappropriately is "seánce." They say: "I want to find a medicine man and get ready for a 'seánce'." What they should say is: "I want to arrange a meeting with a medicine man so that he may give me mushrooms to eat." According to the dictionaries, the word "seánce" means "session or meeting of some group; especially a meeting of spiritualists in which they try to communicate with spirits of the dead."

Among the Mazatecans, the mushrooms have sometimes been eaten in groups; as we know from Wasson's data. But it remains to be proven yet whether it is still the normal custom to eat mushrooms in groups. And even then they do not try to communicate with spirits of the dead, which is the common and almost universal meaning of the word "seánce" for whoever uses it. Neither of these two meanings fit as far as the Mixe use goes, so that such a word needs to be eliminated because it fools the reader by leading him to infer something which the data do not support.

"Narcotic" and "hallucinogenic" have been used as descriptive terms for the mushrooms. There is no inconvenience in using the first term. "Hallucinogenic" can also be justified if we use it only as a description of the effects that are produced in laboratory experiments, or by the mushrooms themselves or their narcotic chemical derivatives. Nonetheless, it is not scientific to apply it to the indigenous use, since this is based on preparatory, prescriptive, rigid and inflexible rules; and so much also on the way in which the mushrooms are eaten, and how one behaves after its ingestion. To apply the term to the indigenous use is to judge, not describe. The indigenous people would not accept such an evaluation. Without trying to give a scientific explanation of the phenomena experienced by those Mixes who ate the mushrooms, it is certain that describing them as hallucinations does not represent a scientific judgment. Such a description does not explain what the caretakers (not having eaten any mushrooms) hear said and see done by those who did eat the mushrooms. Neither does it explain the accurate divination and prophesying done under the influence of the mushrooms, and verified afterward.

To use the mushrooms in the lab would only show the effects of the chemicals in the various participants. Such an experiment would completely lack the effects they could cause in the indigenous eater: the hopes of a cure for his illness, or the solution to his problems, or answers to his questions; the four days' solemn preparation; the isolated, quiet house; and his assimilation into the ritual and spiritual ambience upon eating what to him are "sacred mushrooms."

It is necessary that we further research the *tonalamatl*, the use of the sacred mushrooms, and the relationship between the two.

We know that there is a relationship between the *tonalamatl* and the sacred mushrooms. But we still ignore the extent of this relationship and the significance it has in Mixe culture and thought. Nor do we know the extent to which the *tonalamatl* influences the daily lives of the villages that still use it. If we want to formulate a truthful description from our anthropological studies, instead of a fantasy fabricated by imagination barely related to reality, it is essential that we conduct ongoing research on:

(1) The *tonalamatl*—the meaning of some of the names yet unknown; the significance of the "numbers" that do not look like the ordinal numbers; the significance of the alternate use of the terms "ascend" and "descend" in reference to the division in threes of the *tonalamatl*. Is it still in the habit of some towns (or villages) to determine "the *tona*" of those who are born and how do they do it? According to the day of the *tonalamatl* on which one would be born? Or by asking the mushrooms? Or in some different way? If the day on which a child were born had any influence on the occupation (carpentry, pottery, baking, bonesetting, etc.) he has to follow throughout his life? If being born on a "bad day" means that he will have bad luck in everything and throughout his entire life?

(2) The sacred mushrooms, "the liana" (morning glory), the lily of the valley, and the *toloache*. That which they call "the liana" (*Ipomoea* sp.?), known in Mazatlán and Coatlán, should be studied more. The lily of the valley (*Datura arborea*) and *toloache* (two or more kinds—*Datura meteloides*, *D. stramonium* or *D. sp.*) should also be studied. Which plants are known in which villages? How are they used, and for what? Do they require the same "respect" that is given to the sacred mushrooms? What explanations do the natives give for the extraordinary properties that the mushrooms and other narcotic plants show? Are the other plants for "lay" use, or are they controlled by the medicine men?

(3) *Brujos* and *brujas*: How do they become *brujos* and *brujas*? By eating the mushrooms? By studying or practicing under the training of some who are already *brujos*? Who teaches the men and women? Is it that they voluntarily choose to become *brujos*, or is there someone who nominates them for study?

Once we know all of this information, it is possible that the enigma will have been resolved of the mushrooms and their relationship to the *tonalamatl*, and of the place they occupy in the Mixe culture.

Bibliography

Barradas, José Pérez de. 1957. *Plantas Mágicas Americanas*. Instituto "Bernardino de Sahagún."

Carrasco, Pedro, Walter Miller and Roberto J. Weitlaner. 1959. El Calendario Mixe. *El México Antiguo*. Tomo IX, pp. 153-171. Mexico.

Foster, George M. 1945. Sierra Populuca Folklore and Beliefs. Univ. of California Publications in American Archeology and Ethnology 42: 177-250. Berkeley & Los Angeles: University of California Press.

Miller, Walter S. 1952. Algunos Manuscritos y Libros Mixes en el Museo Nacional. *Tlatoani* 1: 34-35. Mexico, March-April.

———. *Cuentos Mixes*. 1956. Instituto Nacional Indigenista. Mexico.

———, Pedro Carrasco, and Roberto J. Weitlaner. 1959. El Calendario Mixe. (see Carrasco).

Noriega, Raúl. 1954. Claves Matemático-Astronómicas. Summary published by the author for the Roundtable verified by the Mexican Society of Anthropology. Mexico.

Safford, W. E. 1920. Daturas of the Old World and New. Annual Report, pp. 537-567. Washington, DC: Smithsonian Institute.

Weitlaner, Roberto J., Pedro Carrasco, and Walter Miller. 1959. El Calendario Mixe. (see Carrasco).

Wonderly, William L. 1946. Textos en Zoque sobre el concepto del Nagual. *Tlalocan* 2: 97-105. Mexico.

———. 1947. Textos Folklóricos en Zoque. Revista Mexicana de Estudios Antropologicos, Tomo IX, pp. 1-29. Mexico.

Chapter Six
The Hallucinogenic Mushrooms
by Fernando Benítez

Fernando Benítez is a noted professional writer from Mexico, and the following is an essay which originally appeared in 1970 as a chapter of his book, *Los Indios de México.* As we learn from reading it, Benítez was personally referred to the *curandera* María Sabina by R. Gordon Wasson, whom he met during the latter's forays into Mazatec Indian country. Wasson later stated that he "admired the care with which Fernando Benítez followed in our footsteps and, giving us full credit for our discoveries, reported his talks (through Herlinda Martínez Cid, his and our interpreter) with María Sabina" (1980, 122). He added that "Benítez presents an accurate picture of the sacred mushroom complex of the Mazatecs" (1980, 231), and characterizes his account as "rich in fresh insights" (1980, 50). Benítez was one of several speakers featured in 1981 at a roundtable seminar in Mexico City where Wasson was the guest of honor, organized by the Colegio Nacional, as noted by the distinguished mycologist Gastón Guzmán, who also attended (Guzmán 1990, 92).

Unconstrained by formal stylistic conventions of strictly scientific or anthropological investigations, Benítez here gives his sensibilities as a writer free reign, offering a uniquely vivid memoir of his experiences and reflections on the sacred mushroom. His compositional skills hit their stride as he goes into evocative detail about the content and impact of his own encounters with altered perception under the effects of the mushroom. One detects a poignant and refreshing honesty in his account, of a kind not always found in otherwise comparable discussions relating personal impressions of mushroom trips (which on occasion lack cogency, or seem indulgent and shallow). In particular, Benítez' account brings out one inescapable feature of the experience brought on by ingestion of the mushrooms, namely its intensely personal quality. As a result, Benítez powerfully draws the reader into his narrative, offering observations that dovetail with those of authors such as Aldous Huxley, Alan Watts, Huston Smith, and R. Gordon Wasson. He paints an especially compelling portrait of María Sabina, whose essential character seems to emerge as a sort of native Mexican version of St. Theresa of Avila, humbly living a life of complete fidelity to her mystical visions, recognizing in their inspiration an overwhelming reality beyond the reach of doubt. In this respect, his account foreshadows the book *Maria Sabina: Her Life and Chants* by Álvaro Estrada (1981), which first appeared in the original Spanish in 1977. Benítez' contributions have been widely acclaimed. As noted by anthropologist Peter Furst, a leading expert on the indigenous context of hallucinogens, such as the ritual use of peyote by the Huichol (1972, 1976):

> Another non-Huichol who witnessed a peyote hunt in the 1960's, and whose work deserves attention for its wealth of detail and literary quality, is the Mexican writer Fernando Benítez. Although not a trained ethnographer, Benítez is a sensitive observer of Mexican Indian culture. His numerous publications on Cora and Huichol shamanism and on the cultural meanings of peyote and other sacred hallucinogens in indigenous religion and ritual, although intended as literary reportage rather than ethnography, show not only anthropological insight but a rare gift of rendering his observations into literature ... (1972, 144ftn).

The following text contains a rich potpourri of literary, cultural and historic references. Benítez included footnotes in his discussion clarifying some of these, and I have added some additional notes of my own to provide further related information. In the following translation, numeric superscript designates Benítez' notes; those I have added are marked with superscript letters of the alphabet, so as not to disrupt the original numbering of Benítez' notes. By thus preserving the correspondence between the footnote numeration of his original work and the presentation below, this alphabetic notation facilitates comparison between the Spanish and English texts, while also permitting the reader to distinguish instantly between these two groups of notes. (For convenience I have introduced one minor change in the numeration from the original, by designating Benítez' notes a single series, numbered 1-20. In the original text they comprised two series, the first with eight notes numbered 1-8, and the second starting over with twelve notes numbered 1-12. As presented here, only notes 1-8 are numbered true to the original. Notes 9-20 are Benítez' notes 1-12, his second series.)

Benítez includes some fairly substantial quotations from the Wassons' writings. I have referred directly to the original works where possible for these, rather than translating them back into English from the Spanish versions that appear in Benítez' text.

Benítez opens his discussion by referring to *Historia General de las Cosas de Nueva España*, a massive sixteenth century account of Mexican culture and the Conquest era compiled by the Franciscan friar Bernardino de Sahagún. The original historic document includes two codices, known as the Florentine Codex and the Magliabecchiano Codex. The Wassons state that Sahagún, "by reason of his moral and intellectual qualities towers over all his contemporaries who were writing about Mexico in the sixteenth century," adding that his work represented a "systematic and sympathetic study of the Indians among whom he lived and labored" (Wasson and Wasson 1957, 222).

*   *   *

Benítez, Fernando. 1970. Los hongos alucinantes. In *Los Indios de México, Libro I, Tierra de brujos, III*, pp. 205-282. México DF: Biblioteca ERA.

DOCUMENTS

Sahagún tells us that the first thing the Indians ate in their celebrations were certain small, black mushrooms they called *nanacatl* which are intoxicating, causing them to see visions and even arousing lust. They ate them with honey, and when they began to feel their effects (*a calentar*) they danced, they sang, or they cried; some did not want to sing but sat in their rooms as though in thought (*como pensativos*). They saw in a vision that they died, that they were devoured by some fierce beast or that they took captives in war. Others saw in a vision that they would become rich and own many slaves; others saw that they would steal or commit adultery and for this reason would have their heads crushed flat (*de hacer tortilla la cabeza*; literally, "make a tortilla of their head") others saw in a vision that they would kill someone and for this reason would have to be made dead (*habían de ser muertos*); others, that they would live and die in peace; others, that they would drown, they would fall from a height and die of the fall, or in some whirlpool. All the disastrous events that will happen, Sahagún concludes, they saw in a vision. When the intoxication of the little mushrooms (*honguillos*) had passed they spoke with one another (*los unos con los otros*) about the visions they had seen.

In Book X of his *Historia General de las Cosas de Nueva España*, the friar returns to the subject:

> They had great knowledge of herbs and roots and knew their qualities and virtues; they themselves discovered and first used the root known as *peyotl*; and those who took it used it in place of wine. And they did likewise with what they referred to as *nanacatl*; which are the evil mushrooms (*los hongos malos*) that also intoxicate like wine; and they would assemble on a flat plane after having eaten them, where they would dance and sing all night and day as they pleased; and this the first day, and the next day they all cried greatly and would say: they were cleansing and washing their eyes and faces with their tears.

And again in Book XI, he adds these valuable details about the mushrooms:

> Those who eat them ... feel faintness of the heart (*bascas del corazón*) and see visions sometimes frightening and other times laughable; they arouse lust in those who eat many or even a few (*a los que muchos de ellos provocan a lujuria y aunque sean pocos*). And of crazy or mischievous young men it is said they have eaten *nanacatl*.

On the other hand, Francisco Hernández, the physician of Phillip II, left us this most interesting note in his *Historia Plantarum Novœ Hispaniœ*:

> Others (mushrooms) when eaten cause not death but a madness, sometimes lasting, whose symptom is a kind of uncontrollable laughter. They are commonly called *teyhuinti*. They have a tawny color (*color de leonado*), a bitter taste and a certain freshness that is not unpleasant. Others still, being awesome and frightening (*siendo terribles y espantables*), were sought by the same nobility for their banquets and celebrations, reaching an extremely high price, and they were concealed with much care; this species has a dark color and a certain bitterness.[A]

The descriptions of Sahagún and Hernández, so remarkable, offer a satanic perspective, but not directly associated with the devil. It is the vehement one Motolinía[B] who identifies them with the devil, seeing in the native ritual of eating the sacred mushrooms a ceremony similar to the rite of the Christian communion. He said:

> They had another way of getting drunk that made them crueler: it was with certain fungi or small mushrooms (*unos hongos o setas pequeñas*), of which there are such in this land as in Castile; but those of this land are of such quality that eaten raw and being bitter, they drink after taking them and eat them with a little bee honey; and in a little while they see a thousand visions and especially snakes; and as they go out of their senses, their legs and body appear full of worms eating them alive and thus half raving they leave the house wanting someone to kill them; and with this bestial drunkenness and thing they felt (*trabajo que sentían*), it sometimes happened they would hang themselves and they were also crueler toward others. With these mushrooms, called *teunanacatlh* in their language, which means the flesh of the god or the devil which they worship, and in this way with this bitter delicacy they communed with their cruel god.

Communion. Not with God but with the Devil, that terribly active Devil who impregnates the chronicles with his smell and always shows his horns and tail behind all the events. How we recognize the prose and spirit of the sixteenth century in those fragments! Other than the vision of future wealth and a peaceful death, the informants of Sahagún or Motolinía did not communicate any beautiful hallucination; and if they communicated it, the friars kept much of it to themselves in their writings.

We cannot affirm that either intentionally distorted his version. This vision is authentic, but limited; it offers only one half of the truth, the descent to the underworld, death, misfortune, the liberation of malignant instincts, the whirlpool that drags under and drowns, madness and laughter, but even the laughter is a convulsive laughter and of a demonic nature. The other half of the visions, the one that refers to the mystical ascent or the seductiveness of certain images, they silence or hide because in the sixteenth century everything is observed with a moral purpose and everything has a didactic, exemplary sense. The world of the

Indians is the world of darkness and the Devil, just as the world of the conquerors is the world of light and of the true God. This God is alive, as the Devil is alive; both fight incessantly driven to annihilation and the religious chroniclers, as laymen—we recall Juan Suárez of Peralta and Baltasar Dorantes of Carranza—have the duty in this fight of helping God who neither gives quarter nor asks for it.[C]

For that reason the anthropologist and the friar always go hand in hand. They described the mushrooms and their effects with rigor, sparing no details, but they are unable to evade the fundamental consideration that those mushrooms not only belonged to the rites of the conquered, but were in a certain form the flesh and blood of the Devil shared with them—a way of taking the Devil into the body—just as Christians share the flesh and blood of Christ represented in sacred form.

Therefore, the Spaniards preserve the old cultures and at the same time prohibit them without mercy and condemn the idols, temples, codices, and magical drugs to mass destruction, because everything was associated with the demon, and everything belonged to that world of darkness which had to be annihilated in order to create upon its ruins the world of light, of purity and the conquerors' own truth.

Nevertheless, the Colonial period demonstrates that it is much easier to win the bodies of the conquered ones than the souls. The Indians were reduced to slavery without great difficulties, but they continued to be encouraged by their idols, hidden within Christian altars at times, and the mushrooms and peyote continued being consumed by thousands of witches (*hechiceros*) and *brujos* in the seclusion of their remote mountains, despite the efforts of the clergy and the aid lent to them by the Holy Office.

All this seemed buried in forgetfulness. References to the mushrooms cease in 1726, and although the texts of the chronicles were known to some scholars of our century, they were not an object of study nor were they related to the fact that still, they were used in some places of Mexico. The idols had lost their nature as gods and began to live a second spiritual life in art; the magic drugs, in spite of their use, continued to be despised and feared as if the condemnation of the sixteenth century rested upon them, and it was not until Antonin Artaud and Aldous Huxley[D] initiated the demand for peyote from outside, that our country began to be interested in the Indian drugs.

The hour of the mushrooms had not yet arrived. In 1936, the engineer Robert Weitlaner[E] had given a report on certain species of hallucinogenic mushrooms that were consumed in the Sierra Mazateca, and two years later, in 1938, the ethnologist Jean Bassett Johnson wrote an article published in Sweden about a ceremonial ritual of hallucinogenic mushrooms. These two works, destined for the specialists, passed unnoticed and the glory of the discovery and its popularization would have to fall to a banker from New York named R. Gordon Wasson and to his wife, the doctor Valentina Pavlovna Wasson, creators of a new science: ethnomycology.

Summarizing his work, Roger Heim, Director of the Paris Museum of Natural History, wrote:

In 1953, when the two ethnologists from New York arrived in Mexico, his contribution to the ethnomycological chapter was already remarkable although unpublished. The investigations of Mr. Wasson were applied to an analysis of the relations 'between mushrooms and man through his traditions, culinary habits, literature, religion, expressive arts, symbolism and history.' They have opened a road unknown, and explored lands still virgin, which the old geographers in their maps, for want of something better, dressed up with the famous inscription *Hic Sunt Leones* ("Here there be lions," Georges Becker). They have pursued the relations between man and mushroom in all the sources and illustrated them with all possible arguments of linguistic, historical, and psychological nature, which explain the mycophobia of Anglo-Saxons, and the mycophilia of the Slavs, the Provençal, and the Catalonians. In particular the study of primitive Siberian tribes led them to interpret the use of the fly-killer *Amanita (el empleo del amanita mata-moscas)*[F] by those populations as an intermediary of some sort between God and men. At the same time they have confirmed such practices that they investigated in the symbols of Chinese art, in the midst of European towns or in other places by means of comparative examination of their languages and customs, in the form the mushrooms were able to be used in the first ages of those civilizations. A thesis about the role of those demonic beings in psychogenic manifestations of the towns was derived little by little, supported by a multitude of new or rediscovered data. An original theory on the history of religions was introduced. Wasson thus discovered the survival of certain ancient practices of a similar nature in New Guinea, in Borneo and Peru. But it was Mexico that offered him an exceptional mine of documents in this respect. The remarkable work published in two volumes in 1957, *Mushrooms, Russia and History*, constitutes a monumental contribution to those diverse facets of a new science. The ethnological and linguistic aspects of the Mexican mushrooms were already dealt with generously in two chapters of that work, and outlined a daring but exciting opinion: one that applied by an extension of practices in native Siberia to phases found again in Borneo, New Guinea, Peru, and Mexico, following the routes of migration established according to the opinion of certain ethnologist.[1]

At the beginning of 1953, Wasson, already knowing the works of Weitlaner and Johnson, had knowledge that an American linguist, Eunice Victoria Pike, had lived a long time in the Sierra Mazateca and he wrote to her requesting information on the hallucinogenic mushrooms.

Her answer is a remarkable ethnographic document that in certain ways picks up from the investigations undertaken by the friars and naturalists of the sixteenth century. And like her remote predecessors, Miss Pike was not only a linguist and compiler of facts concerning the Indians, but by her character as a Christian missionary and witness, observed the survival of the sacred mushrooms with manifest displeasure.

In addition, everything Miss Pike knew about the mushrooms she knew through informants not directly, for this strict Protestant had never been allowed to attend a ceremony much less to share in those dark vegetal demons. By any standard, her letter[2] offered a perspective capable of driving the most objective investigator crazy and Wasson decided to explore for himself the remote and almost forgotten Sierra Mazateca.[G]

Wasson went in answer to a call to give fame to the *nanacatl* of the Indians. There was no one in the world better prepared nor who felt more passion for that vast, fragile, delicate and mysterious universe of the mushrooms. Like all discoverers, he had to remove them from the darkness, and at the same time contribute to their annihilation by dissipating the atmosphere of love and reverence that had surrounded them until then.

The Discovery of the Mushrooms

Wasson departed from Mexico City on August 8, 1953, in the company of his wife, Dr. Valentina Pavlovna Wasson, their sixteen-year old daughter Masha, and the engineer Robert Weitlaner. They spent the night in Teotitlán, the ancient city of the gods, and from there they began the climb up the Sierra. At that time the path (*la brecha*) leading to Huautla did not exist. They went by the paths of *recuas* (*recua*: a multitude of things in succession), mounted on a "miserably small and thin" horse and five mules under the charge of a carrier, a Mazatec Indian named Victor Hernández. That night they arrived in Huautla and stayed over in the house of the teacher Herlinda Martínez Cid, a friend of Miss Pike. Herlinda could do nothing other than introduce them to Aurelio Carreras, a one-eyed Indian, forty-five years old, owner of two or three houses and vaguely connected with the mushrooms. Wasson went completely lacking information. Those who went knowing—a relative of Herlinda; the priest of Huautla; and Concepción, the wife of a lost, alcoholic *curandero*—tried to help him, but apparently no one had any great knowledge about the mushrooms.

At night, the one-eyed Aurelio, Concepción and Victor, the carrier, brought mushrooms wrapped in banana leaves or pieces of fabric and Aurelio advised silence because this was a "very delicate" matter.

Describing the atmosphere of mystery that still surrounded the mushrooms, Wasson says:

> When approaching the Indians with our inquiries we were careful to speak of the mushrooms with the deepest respect. (After all, it was a bold thing that we were doing, strangers probing the innermost religious secrets of this remote people. How would a Christian priest receive a pagan's request for samples of the Host?) (Wasson and Wasson 1957, 250).

Wasson did not hesitate to ask diverse people for explanations about the mysterious power of the mushrooms. One source told him: "Our Lord crossed the countryside and wherever he spat there a mushroom grew." ("I think," Wasson writes, "that to spit is a euphemism for ejaculate.") One woman confided to him that *'nti¹ ši³ to³* meant "sprouts from the blood of Christ that Mary was unable to recover."[H] (Wasson notes: "To me this recalls the observations of Miss Pike.") And the same woman added that the *'nti¹ ni³ se³⁻⁴*, the smallest of the mushrooms, "appeared wherever Christ staggered under the weight of the Cross."[3]

Aurelio was most explicit. According to him, the mushroom "is speech" and it speaks of many things: of God, of origins, of life and death, and it says where to find lost objects. It shows where God is.

With all this, Wasson was filling his notebook with interesting notes:

> We learned that the Mazatecs are mycophagous and many kinds of edible mushrooms are offered on market day in the plaza. Each has its own name and the general word for 'mushroom' is tai^3, the t being explosive and each of the vowels being nasalized. But this word embraces only mushrooms other than the sacred ones. Each of the sacred kinds has its own name, and all together they are called $ši^3to^3$. This name is invariably preceded by another verbal element so that the normal expression is, as Miss Pike had told us, $'nti^1 ši^3 to^3$, the first syllable conveying a sense of deference and affection. (The apostrophe represents a glottal stop.) The word $ši^3 to^3$ means literally 'that which springs forth' ... a felicitous mystical metaphor ... the word is saturated with *mana*: it is uttered in a whisper and Victor was loathe to pronounce it at all; when he had to use it, he would substitute a gesture—his gathered fingers making an eating motion before his mouth (Wasson and Wasson 1957, 251).

The sacred mushrooms are never sold in the plaza of the market, although all the accessories of the rite can be bought there without difficulty.[4]

The Divinatory Powers of the One-Eyed Aurelio

The time arranged for by Wasson drew toward its end and he had managed neither to make contact with a *curandero* or $čo^4ta^4ni^4če^4$, nor to attend a mushroom ceremony. Don Roberto had a hunch: what if Aurelio was the *curandero* that they were seeking? At that moment Aurelio made his silent appearance. "And tell me, Aurelio," asked Don Roberto, "are your experiences always successful?" "Yes, always," replied Aurelio. "A son of Mr. Wasson's is in Boston and he desires to have word of him. Could you help us tonight?"

We should clarify that in 1953 the mushrooms were not utilized for the sole purpose of provoking ecstasy. Whenever they were employed it was to cure a disease or resolve a problem, and whoever contracted with a *curandero* had to present a definite matter for his consideration. Wasson, in fact, had not received any letter from his son Peter, a boy eighteen years old who worked in a company in Boston, and although he was preoccupied, he presented his problem, not because he believed in Aurelio's divinatory gifts, but as a pretext to attend the desired ceremony of the hallucinogenic mushrooms.

Aurelio consulted with the teacher Herlinda and after many deliberations agreed to the request of Don Roberto. They would have to be ready at nine o'clock that evening and were advised there were different ways to carry out the ceremony. Some *curanderos* chanted, sang, or even shouted. As the mushrooms express themselves only in Mazatec, he requested that his son Demetrio accompany them in order to translate the words of the mushroom.

The ceremony, celebrated in Aurelio's ramshackle house, was the classic ceremony in which the Mazatec *curanderos* traditionally officiate: an altar with

Catholic saints, candles, eggs of fowl (*totola*[1]), *copal, pisiete*, feathers of macaw (*guacamaya*), *cañutos* of *aguardiente*[1] and a roll of bark paper.

At eleven o'five Aurelio asks: "Where is Peter?" In Boston. Aurelio puts out the candles and two hours pass in absolute darkness and silence. At one o'five a storm comes over Huautla. Thunder, lightning, rain. Abruptly a gunshot sounds. Demetrio exclaims: "Murder." Bare feet run along the footpath near the house. A door is beaten upon. Three more gunshots are heard. "Throughout the storm and the shooting Aurelio proceeded deliberately with the ritual" (Wasson and Wasson 1957, 257). It is difficult to distinguish Peter because he is far away in a foreign city.

Later, the mushrooms declare their sentence: "Peter is alive but 'they' are reaching out for him to send him to war. Possibly 'they' won't get him, but it is hard to say. Germany seems to enter into the situation" (Wasson and Wasson 1957, 261). Subsequently the mushrooms affirm that Peter is not in Boston as Mr. Wasson believes but in New York. Great difficulties are causing him to almost lose his head and he is thinking about his family to the point of crying. He has never had such problems. He does not know how to explain to his parents what is happening to him.

Before concluding the ceremony at one forty-five, the mushrooms reveal that one of Wasson's relatives "is destined to fall seriously ill within the year." Out of this extensive ceremony, Wasson recalls the serious look that Aurelio Carreras' single eye directed toward him.

Wasson writes:

Here we should prefer to bring our story of Huautla to an end but candor compels us to add a few more lines. Our attitude toward the divinatory performance and especially the oracular utterances had been one of kindly condescension. We said to ourselves that it was cruel on our part to ask Aurelio, locked up in his unlettered Indian world, to enter understandingly into the problems of the Wassons of New York. His divinatory powers, put to this appalling test, had seemed to us pitifully thin, but of course we had duly entered in our notes all that he had said ...

We reached home in the second week of September. In the kitchen of our New York apartment we found the leavings of a party that during our absence Peter and his friends had held. The bills from the purveyors supplied the date: the weekend of August 15-16! Peter easily confirmed this when we saw him. Laughingly we credited the sacred mushrooms with a hit, a palpable hit, and then gave the matter no more thought.

Aurelio's prediction about the army had seemed badly aimed. After all, Peter at the age of 17 had enlisted in the National Guard, and this gave him exemption from the draft. Soon after our return to New York, RGW left for Europe on a business trip, and late in the morning of Monday, October 3, he arrived at Geneva. There a cablegram from home awaited him with sensational news: Peter had just made known his settled determination to enlist in the regular army for a three year term. He had come to this decision after a prolonged emotional crisis involving a girl, and that crisis ... had been boiling while we were in Mexico. Would RGW please send a cablegram at once begging Peter to postpone his rash step? RGW sent the message but, before it reached Peter, he had signed up ... days passed before Aurelio's pronouncements suddenly came to mind.

A few months later, after the usual training period, the army sent Peter abroad for duty, but to Japan not Germany.

There remained one final prediction: grave illness was to strike the Wasson family within the twelvemonth. (In the Mazatec world the 'family' embraces all the kin.) This seemed on its face unlikely, for our families are unusually small ... In January (Benítez states February 1954—Ed.) one of RGW's first cousins ... abounding in vitality, suddenly died from heart failure (Wasson and Wasson 1957, 264-265).

An Encounter with María Sabina

In 1955 Wasson returned to the Sierra, this time accompanied by his friend the photographer Allan Richardson. Wasson felt isolated. Aurelio, the remarkable diviner, was ill; Concepción did not dare to place upon his shoulders the responsibility for officiating for the two foreigners and even the same priest was absent from Huautla. There were therefore no hallucinogenic fungi, no willing *curanderos*, nor hopes for celebrating a ceremony.

Wasson—some Mazatec god must have inspired him with his bold determination—then went to the city council and there found seated at a large table the Syndic (*síndico*),[K] Cayetano García, a young Indian man of thirty-five who spoke Spanish. "May I speak confidentially with you?" Wasson asked. "Of course," replied the Syndic, "tell me what it concerns."

I can imagine without any effort Wasson at that critical moment of his career, leaning toward Cayetano's ear and telling him in his soft voice and with utmost courtesy mingled with humor: "Would you help me to learn the secrets of '*nti¹ ši³ to³*?'"

The proposal, thus formulated, was all the more amazing considering he had only just arrived, and according to his own admission without hiding his satisfaction, he pronounced correctly the feared Mazatec word "with the glottal stop and the tonal differentiation of the syllables." Cayetano opened his eyes admiring what he heard. "Nothing would be easier," he replied. "Please come to my house at *siesta* time."

Cayetano lived at the edge of Huautla, in a two-story house with one side facing the main street and the other side the lofty slopes of the mountains. A house, like most of the mountains, full of men, animals, teeming with life. "A hen," writes Wasson, "sitting on her eggs on one of the cluttered tables was a silent witness to all that went on" (Wasson and Wasson 1957, 288).

Having barely arrived, Cayetano and his brother Genaro had them descend the mountain to a rudimentary mill, where on the sugar cane *bagasse*—oh immortal gods!—they discovered an enormous supply of the mushroom called *ki³šo¹*, landslide (*desbarrancadera*). Wasson did not hide his emotion:

> We photographed them to our hearts' content. We gathered them in a pasteboard box: the mushrooms must always be carried in a closed parcel, never exposed to the view of passers-by. They were a noble lot, mostly young, all of

them perfect in their moist health and fragrance. Then we carried them up the steep mountain side to the house. We were warned that if we saw any dead animal on the way, the mushrooms would lose their virtue—happily, we saw none (Wasson and Wasson 1957, 288).

Wasson had managed to penetrate the wall of fear and distrust with which the Indians tend to defend themselves around strangers. Cayetano, in spite of the fact that he spoke Spanish and held a position on the city council, was a man strongly rooted in his native soil, who believed in the sacred power of the mushrooms and turned to them to resolve many and serious problems that Mazatec families always face. His feeling of solidarity, so characteristic of the Indians, his knowledge of the magical life that had developed in Huautla outside of intrusive snooping, changed the situation for Wasson. He was not satisfied by merely showing him the fungus *in situ*, and sent Wasson on a hike with Emilio, his other brother, to a certain place where they would find a "first rate *curandera*."

The *curandera*—is it necessary to explain that it was María Sabina?—was seated on a *petate*, with one of her daughters, and did not realize that along with this foreigner—the Fat Joker to some Mazatecs—entered recognition of her powers, world renown represented by articles in magazines of wide circulation, books and scientific monographs, disks that would record her shamanistic songs, photographs, movies, and caravans of tourists eager to know the mysteries of the sacred mushroom.

María agreed to officiate that very night in a ceremony—perhaps the first in which the mushrooms were taken for the purpose of inducing ecstasy—and Wasson, without any special aid, in a single afternoon, accomplished in one stroke all his objectives.

Wasson writes:

On that last Wednesday of June, after nightfall, we gathered in the lower chamber of Cayetano's house. In all, at one time or another, there must have been twenty-five persons present, mostly members old and young of Cayetano's family ... Allan Richardson and RGW were deeply impressed by the mood of the gathering. We were received and the night's events unrolled in an atmosphere of simple friendliness that reminded us of the agape of early Christian times. There was no familiarity (Wasson and Wasson 1957, 289).

We were attending as participants a mushroomic Supper of unique anthropological interest, which was being held pursuant to a tradition of unfathomed age, possibly going back to the time when the remote ancestors of our hosts were living in Asia, back perhaps to the very dawn of man's cultural history, when he was discovering the idea of God (Wasson and Wasson 1957, 200).

Who could say? Wasson writes:

At first we saw geometric patterns, angular not circular, in richest colors, such as might adorn carpets or textiles. Then the patterns grew into architectural structures, with colonnades and architraves, patios of regal splendor, the stonework all in brilliant colors, gold and onyx and ebony, all most harmoniously and ingeniously contrived, in richest magnificence extending beyond the reach

of sight, in vistas measureless to man. For some reason these architectural visions seemed oriental, though at every stage RGW pointed out to himself that
they could not be identified with any specific oriental country. They were neither Japanese nor Chinese nor Indian nor Moslem. They seemed to belong
rather to the imaginary architecture described by the visionaries of the Bible. In
the æsthetics of this discovered world attic simplicity had no place: everything
was resplendently rich.

At one point in the pale moonlight the bouquet on the table assumed the
dimension and shape of an imperial conveyance, a triumphal car, drawn by living creatures known only to mythology. With our eyes wide open, the visions
came in endless succession, each growing out of the preceding one. We had the
sensation that the walls of our humble house had vanished, that our untrammeled souls were floating in the universe, stroked by divine breezes, possessed
of a divine mobility that would transport us anywhere on the wings of a
thought. Now it was clear why Don Aurelio in 1953 and others too had told us
that the mushrooms would take you *ahi donde Dios está*—there where God is.
Only when RGW by an act of conscious effort touched the wall of Cayetano's
house, would he be brought back to the confines of the room where we all
were, and this touch with reality seemed to be what precipitated nausea in him.

On that night of June 29-30 we saw no human beings in our visions. On
the night of July 2-3 RGW again took mushrooms in the same room, with the
Señora again serving as votary. If we may anticipate our story, on that second
occasion RGW's visions were different. There were no geometrical patterns, no
edifices of oriental splendor. The patterns were replaced by artistic motifs of
the Elizabethan and Jacobean periods in England—armor worn for fashionable
display, family escutcheons, the carvings of choir stalls and cathedral chairs.
No patina of age hung on them. They were all fresh from God's work-shop,
pristine in their finish. The beholder could only sigh after the skill that would
have fixed those beauteous shapes on paper or in metal or wood, that they
might not be lost in a vision. They too grew one out of the other, the new one
emerging from the center of its predecessor. Here as in the first night the visions seemed freighted with significance ... We felt ourselves in the presence
of the Ideas that Plato had talked about. In saying this let not the reader think
that we are indulging in rhetoric, straining to command his attention by an ex-
travagant figure of speech. For the world our visions were and must remain
'hallucinations.' But for us they were not false or shadowy suggestions of real
things, figments of an unhinged imagination. What we were seeing was, we
knew, the only reality, of which the counterparts of every day are more imper
fect adumbrations. At the same time we ourselves were alive to the novelty of
this our discovery, and astonished by it. Whatever their provenience, the blunt
and startling fact is that our visions were sensed more clearly, were superior in
all their attributes, were more authoritative, for us who were experiencing them,
than what passes for mundane reality.

Following the visions that we have already described, on both occasions
RGW saw landscapes. On Wednesday they were of a vast desert seen from afar
with lofty mountains beyond, terrace above terrace. Camel caravans were advancing across the mountain slopes. On Saturday the landscapes were of the es-
tuaries of immense rivers brimming over with pellucid water, broad sheets of
water overflowing into the reeds that stretched equally far from the shore line.
Here the colors were in pastel shades. The light was good but soft as from a

horizontal sun. On both nights the landscapes responded to the command of the beholder: when a detail interested him, the landscape approached with the speed of light and the detail was made manifest. There seemed to be no birds and no human life in the river estuary, until a rude cabin suddenly appeared with a woman motionless nearby. She was a woman by her figure and face and costume, and of course the vision was in color. But she was a statue in that she stood there without expression, doing nothing, staring into the distance. She might be compared to those archaic Greek sculptures where the woman gazes into space, or, better yet, the departing woman on the Greek funerary stele who looks into eternity ... (Wasson and Wasson 1957, 293-295).

It seemed as though I was viewing a world of which I was not a part and with which I could not hope to establish contact. There I was, poised in space, a disembodied eye, invisible, incorporeal, seeing but not seen (Wasson 1957, 109).

PHOTOGRAPHS and captions (plate between pp. 224 and 225):

1. (the Sierra Mazateca, low altitude aerial view): "Cold land, our land of fog."
2. (closeup, hands picking mushrooms, terrestrial meadow habitat with herbaceous plants): A drop of Christ's blood fell among the plants.
3. (María Sabina) "Flowers that clean while I walk ... "

Pieces of the Puzzle

We must leave Wasson, discoverer and describer of the hallucinogenic mushrooms, but not without adding that his remarkable investigations, far from being restricted to the Sierra Mazateca, soon extended to places as close to the capital as San Pedro Nexapa on the slopes of Popocatépetl and Tenango del Valle in the vicinity of Toluca, or as remote and inaccessible as the Mixeria, the Zapotecs in the coastal sierra, the Chatino country, the Chinantla and the Alta Mixteca. But this investigator went further still and managed to interest a group of scientists in the hallucinogenic mushrooms. From 1956 their explorations no longer included only their family or an occasional scholar, but an entire team of eminent specialists in chemistry, botany, ethnology and linguistics. Dr. Roger Heim, accompanied by his assistant Roger Cailleux, obtained stock cultures and spores from Mexico to produce hallucinogenic mushrooms in the laboratory, analyzing and describing them while they traveled through Mexico and Central America, attended ceremonies, personally experienced their effects and wrote scientific monographs and descriptive articles. Dr. Guy Stresser-Péan, of the Paris Museum of Man, studied diverse ethnological aspects of the $'nti^1$ $\check{s}i^3$ to^3; Dr. Albert Hofmann of Sandoz Laboratories in Basel isolated and synthesized psilocybin; Dr. Aurélio Cerletti with his colleagues studied the physiological and pharmacological properties of the mushroom; members of the Paris Academy of Medicine, under the direction of Dr. Jean Delay, experienced the effects of psilocybin on mentally ill and normal individuals; and the Summer Institute of Linguistics translated vocabulary and texts into five languages. In this fashion, a layman coordinated a scientific collaboration of international reach that

among other accomplishments made possible the publication of that monumental monograph *Les Champignons hallucinogènes du Mexique*, unfortunately not
yet translated into Spanish.

In addition Wasson studied traces the mushrooms had left in Mesoamerican
archeology, salvaged notes and brief references to peculiar facts left in the histo-
ries, the dictionaries and writings by chroniclers famous or obscure, naturalists,
linguists, or simple *aficionados*. It could be said that Wasson exhausted the matter, in his passion even going to the extreme of exploring that vast quarry con-
sisting of the thousand volumes of the Holy Office kept by the *Archivo General
de la Nación* in search of proceedings brought by the Inquisition against the heretical hunger for the sacred mushrooms.

But Wasson had not yet finished his work. The death of his wife and collaborator, Dr. Valentina Pavlovna Wasson, which dealt him a severe blow, and
his professional work as a banker did not prevent him from undertaking exhaustive explorations and accumulating new and important documents.

Thanks to his tenacity, he had the fortune to witness religious ceremonies
that have been kept intact despite persecutions and changes. This has truly been
a unique experience because it has allowed us to know one of the highest and
most significant spiritual manifestations of the Mexican Indians. We seem to
have been furnished with the scattered, almost unrecognizable pieces of a gigantic puzzle which generations of investigators have tried in vain to decipher: the
origins of man in the Americas, the dispersal of a culture originating in Asia
whose tracks are identifiable not in stones, nor in fossils, nor in lasting traces,
but curiously in the fragility of a mushroom, in shamanistic practices, in deliriums and ecstasy amidst Siberian snows or under the leafy trees of the Sierra
Mazateca which speak to us of a spiritual unity, of a nostalgia and a desire that
remain alive in the hearts of men.

Peyote and Hallucinogenic Mushrooms

Until 1957 peyote was considered the unquestionable ruler of the Mexican
hallucinogenic drugs. It had managed to command like no other drug the curiosity of the first chroniclers and describers of our flora in the sixteenth century; it
maintained its hegemony throughout the Viceroyalty, as demonstrated by the
numerous judgments by the Inquisition concerning it; it kept its prestige intact
during the nineteenth century, and in the third decade of our century it began to
claim an international renown. About the hallucinogenic mushrooms, however,
no one spoke. Contained within the zones of the Sierra Mazateca and the Mixeria or consumed in greater secrecy by isolated witches (*hechiceros*) of other
regions, their worship and marvelous properties were known only vaguely
within a restricted group of scholars and linguists, but in less than six years *teonanácatl*, the food of the gods, began a dizzying rise and yielded a family tree
and the abundant credentialed documents that figure in Wasson's bibliography
of the hallucinogenic mushrooms of Mexico and psilocybin, published at the end
of 1962 (Ed.—1963) by Harvard University.[5]

Now peyote and the mushrooms are of equal standing and travel throughout Mexico by parallel roads within their native land. Although of different nature they are twin drugs, gods and demons at the same time, objects of reverence and fear marked by a common destiny, by identical rituals and spells, by characteristics, adventures and mysteries so similar between the two that it is often difficult to distinguish them.

Affinities of the Twins

First we note that the *curandero*, to commune with these gods, undergoes a transformation by which he himself becomes a god. No effort is necessary to demonstrate the existence of these gods. To this day in Huautla proof that the mushroom is sacred is established by the incontrovertible fact that eating it is sufficient to feel its supernatural effects.

Without these precious gifts of Nature the *curandero*, apart from obligatory exceptions, is a man deprived of power. It is not possible for him to undertake mystical ascent, nor descent to the underworld of the dead, nor divine the causes of illness, nor predict the future. The deities of these drugs speak for themselves, they act directly upon the *curandero* permitting him to establish communication with another class of divinities. The persecutions suffered, the edicts directed toward their destruction, and the idea of seeing a devil hidden in the mushrooms and the cactuses, determined that their followers endowed them with another personality which they added to their well-proven divine forces, the force and prestige of the Christian divinity, which made of them a formidable concentration of magical and sacred powers.

"In native medicine," writes Aguirre Beltrán, "the medicament, the rite and the incantation are essential elements of the magical practice; but the accent put upon that which is used carries such emphasis, that what is done and what it said rises to a plane of secondary importance."[6] This hierarchical organization, applying to the Colonial period, has lost nothing of its validity. Cactuses and mushrooms are harvested at dawn on propitious days and consumed by night taking advantage of the silence and darkness. Their handling requires complicated acts of purification involving roles for tobacco, incense burning, sexual abstention, and sometimes fire. The banquets are celebrated before altars previously swept and decorated with flowers, and the patient as well as the *curandero* must maintain the same state of purity.

In the Huichol altar or the Mazatec family room where the altar is set up and the ceremony occurs, the dish containing the peyote or mushrooms is emphasized as a tabernacle for safekeeping. Speech is in a low voice and all eyes are fixed upon the little vegetal creatures (*las pequeñas criaturas vegetales*) who are going to perform the miracle of giving that group of men, frequently miserable as beggars, the omnipotence and omniscience of the gods.

The Identity of the Opposites

Both drugs are administered "married" or in pairs, a masculine specimen accompanied by another feminine, and they are customarily eaten with chocolate or sugar to disguise their bitterness and "facilitate the liberation of the alkaloids."[7] Nor do the surprising affinities between peyote and the mushrooms end here. When the *curandero* deals with them he uses the language of the divinity; the metaphors, the forms and reverential diminutives used in naming them figure among the most beautiful expressions of the Indians.

Their effects are equally similar. In essence, they both cause sensory hallucination, a dissociation (*desdoblemiento*) of the personality, alteration of time and space, inability to pay attention, reminiscences, and periodic hilarity. This medium of dreams, of deliriums and ecstasy was and is taken advantage of for the purpose of discovering the cause of an illness, the place where a patient lost his soul, or divination of the future. Nevertheless, peyote and the mushrooms cannot be seen from that narrow a perspective. In the magical world of the Indians, suffering is caused by the anger of the gods: *Qualani in Huehuetzin*, "the Ancient God is angry;" *qualani in Chicomecoatl*, "Seven Snake is angry;" *qualani in Chalchiuhtlicue*, "She of the Jade Skirt is angry;" replies the medic consulted when interrogating himself about the cause of pain, Aguirre Beltran writes.[8]

It is therefore indispensable to determine the deity responsible for an evil, to favor it by means of offerings and prayers and to organize the defense of the patient by mobilizing the resources the *curandero* has.

The benefits that the Indians have obtained from their hallucinogens are incalculable. Confronted with a hostile milieu, subjugated to slavery and the plunder of their wealth, stalked by a thousand dangers, Indians, blacks and mestizos have resorted to peyote, to the mushrooms, to *ololiuhqui*, to the Crushed Green (*Verde Machacado*) and to *Señor Estafiate*[L] as the only way to resolve their problems and to alleviate the anguish that dominates them, not by means of the voluntary manufacture of artificial paradises, since it is difficult and perilous to commune with a god and to become a god. They are not given the omniscience and omnipresence of the gods for free. There is a price to be paid for obtaining a state of grace that permits us to communicate with the divinity and to transcend our human condition. This price is abstinence, the purification of the soul and of the body, and on not a few occasions, pain and derangement (*desgarramiento*). These alone are the conditions to divine the hidden causes of our sufferings, to dissociate our personality by means of the most strange and peregrine metamorphosis and free us from the burden, increasingly heavy, of our anguishes and frustrations.

In this fashion, the divergent lines of magical medicine and rationalist medicine are joined, overcoming the abysses that separated them. The magical drugs and the resources put into play by the *curandero* enter into the present day unsuspected as the anguish modern man suffers from worsens. Unfortunately, we do not have anything similar to the figure of the *curandero* who, due to his deep in depth knowledge of the spirit and nature as well as his exceptional virtues, was the one in charge not only of reducing his own anxiety but to offer security and consistency for the group entrusted to his hands. The *curandero* has in a certain way been the Moses who removed his town from slavery and gave them strength to undergo the terrible tests lowered upon them in their long wandering towards a better life. Playing his historic role, the *curandero* is on the verge of extinction—we attended the final collapse of the postulates upon which his life rested—but we have left the way shown by their magical drugs and, mainly, we have left as an example the ancient method whereby those masters tried to alleviate the anguish, disintegration and insecurity of the human soul.

MARÍA SABINA AND HER SHAMANIC SONGS

María Sabina is an extraordinary woman. As with other notable Mexicans, recognition has come to her not from her mother country, but from foreigners. Roger Heim speaks of the "powerful personality" of María Sabina, and Gordon Wasson, her discoverer, calls her the Señora and writes of his first encounter with her: "The Señora is at the height of her power and it is easy to understand why Guadelupe[9] told us she was a lady without blemish, immaculate, because she alone had managed to save her children from all the fear sicknesses that descend upon children in Mazatec country, and she had never dishonored her power by using it with malicious intent…we have verified that this is a matter of a woman of rare moral and spiritual elevation who has devoted herself to her vocation, and an artist who dominates the techniques of her position. She is truly a personality."

Unfortunately, the fact that María speaks Mazatec exclusively has prevented me from knowing her in all her spiritual wealth and depth. Not without overcoming an old distrust, she agreed to tell me her life story in three sessions, and although she had as translator the intelligent teacher Herlinda, a native of Huautla who speaks Mazatec perfectly, it was quickly revealed that not only was she incapable of translating María's poetic thought, but also distorted the meaning and originality of her story in passing it through the filter of another culture and sensitivity.

Accompanied by her little grandson (*nietecito*), María Sabina always descended the hill against which the hotel leans, which gave me the impression she came flying down from her remote cabin. She literally descended the tile roof, disdaining the door and stairs, and as her bare feet made not the least noise treading on the boards of the runner and she appeared suddenly, without being announced, in an entirely ghostly way, she never let me be surprised when she said near my ear in a very soft voice: "Dali."

Her great-grandfather Pedro Feliciano, her grandfather Juan Feliciano and her father Santos Feliciano were all *curanderos*. She knew none of the three; her father disappeared young, when María was four years old, so she could not take advantage of the knowledge and experiences of her ancestors.

Her family remained very poor and as a girl María Sabina, with her older sister María Ana, had to herd a flock of goats. Hunger drove them to look for the many mushrooms that grow on the hillsides and they ate them raw, common kinds and hallucinogenic. Intoxicated, the two girls knelt down crying and asked the sun to help them.

María, leaving the chair in which she is seated, kneels in the middle of the room and with joined hands begins to pray fervently. She realizes that words are insufficient and resorts to action so that I have a precise idea of the state of religious inspiration the mushrooms plunged her into and what her encounter with them signified. Her expressive face is illuminated reflecting the mysterious light of that first intoxication so distant in time and still so alive in her memory.

"Why are you crying?" I asked her. "It is crying of emotion (*lloraba de sentimiento*). To think about her misery and her desolation." "From then on she ate the mushrooms frequently?" "Yes. The mushrooms gave her courage to grow, to fight, to endure the pains of life."

By the age of six or seven she had already cultivated her father's land with a grub hoe (*azadón*), spun the cotton, and woven her own *huipiles*. Later she learned to embroider, carried firewood and water, sold fabrics or exchanged them for hens, helped grind corn and hunt for mushrooms and herbs in the countryside, that is to say, she worked like all Indian children rising before dawn and not resting a bit until the hour of going to bed.

At the age of fourteen she was proposed to for marriage by Serapio Martínez, a walking merchant who traveled to Tecomavaca, Tehuacán, Cordova, and Orizaba, carrying pots, clothes and a blanket. On one of those trips they took him to fight the *carrancistas*[M] or the Zapatistas, he does not know which, and he returned after eight months, later slinging cartridge holders, bringing horse and rifle, because he was a brave soldier. María told him: lay down your weapons. I suffer a great deal and need you to live with me.

Serapio deserted. He walked trading for some time and visited her secretly. Never, in his times as a retailer or soldier, did he forget to send her some money. María, for her part, continued working and helping with the expenses of the household.

The union—Indians did not marry then—lasted six years. Serapio contracted the Spanish influenza and agonized for ten days on a thrown down *petate*. In vain they went to the best *curanderos* in Huautla. The boy "was like a crazy person" and two days before dying, the *brujos* pronounced: "There is no remedy for him. You will lose your husband."

When the forty days of the official Mazatec mourning passed, María again cultivated the soil and took care of the three children from her marriage: Catarino, María Herlinda and María Polonia. Naturally she ate the mushrooms that gave her the comfort and strength to provide for her children. She lived as a widow thirteen years, cutting coffee on the farms, embroidering *huipiles*, and carrying out small business. From time to time she resorted to the mushrooms, but as her life improved and her children grew, she ended up forgetting about them. She concluded that long period of solitude—"Here we live like nuns," explains the teacher Herlinda—when a man named Marcial Calvo asked for her, a *brujo* by profession, and she had six children with him.

"What distinguishes a *brujo* such as Marcial from a *curandera* like María?" I ask her through Herlinda. "I divine," replied María excitedly. "I arrive at a place where there are dead persons and if I see the patient laid out and hear gentle crying, I feel that a pain approaches. Other times, I see gardens and children and I feel the sick one will recover and the misfortunes go away. By singing I divine everything that is going to happen. The *brujo* scares away the bad spirits with prayers, and the priest by means of offerings. I never ate mushrooms during the twelve years our marriage lasted because he went to bed with me and since he had another way to cure, I always hid my 'science' from him."

Apart from being a *brujo*, Marcial was a bad man. He was accustomed to drinking *aguardiente* as a practice associated with his profession, and it had made a drunk of him. He did not bring in money, and would hit his children and woman even though she was pregnant. From María's story a word arose frequently that I had already heard many times from the mouths of Indians: suffering. "I have suffered a lot, I have suffered too much," she said summarizing the different stages of her life.

Her initiation into magical medicine took place during the final years of their marriage, when two of his well-known elders became ill and according to custom turned to Marcial's professional services. It was worth nothing but eggs, herbs, and prayers. They got worse each day and would have died if María had not intervened to restore their health. How were they healed? "By eating mushrooms. By singing. Invoking the Holy Spirit of God, of St. Peter, St. Paul, and all the saints in heaven."

When Marcial discovered that María ate mushrooms and was a gifted *curandera* with powers superior to his own, he became angry and hit his woman in front of the elders. "Most Holy Mary, I am bleeding!" she exclaimed, her eyes flashing with rage.

"It was tiring, very tiring." Marcial's brutality determined that little by little she "rejected him" according to Herlinda's account. Marcial then "cheated" *("se metió")* with a certain married woman, a neighbor of María who had large children, and one night her husband and the children beat him in the head with clubs. María heard shouting. Nevertheless, she did not think about Marcial until the next day when he was found dead in the road. The deceived husband, with his children, abandoned the adulteress and to this day she lives alone in Dry Gulley (*Barranca Seca*).

The Book of *Sabiduría* (shamanistic knowledge)

Twenty years ago the *brujo* Marcial died. For twenty years María has lived intensely dedicated to the dual task of acquiring a reputation as a *čo⁴ta⁴ni⁴če⁴*, "one who knows," and providing for her more and more numerous family. At first things were difficult. She had to look after her ten children—seven of them are still alive—and her sister María Ana, helping her with the grub hoe, embroidery, the pigs and hens or by selling *aguardiente* and food to the visitors who travel by the dirt road where she has always had her house.

She passed her long period of widowhood alone, not because she thought badly of men, but because having so many children she did not want to get married again and once she began to work with the mushrooms, men lost interest in her.

Her first patients were elders who were dying. Healing them opened up a new avenue for her, and although she had not lost faith in *curanderos* she was afraid to cure using the sacred mushrooms. What resolved her to employ them again was a serious condition she saw in her sister María Ana. While seated or eating, suddenly "she was stricken" (*de pronto "se ponía morada"*), tightening her hands and falling to the floor. The *brujos* had exhausted her with their remedies and María thought that by taking a large quantity of mushrooms she would be able to see the illness and cure it.

On that occasion she took thirty pairs, and while under their effects a spirit with a book in hand approached her and told her: "I am giving you this book so that you can work."

She was unable to read the book, because she had not had the opportunity to go to school, but the gift of divining the future and secret things "as if reading it in a book" was given to her. Due to her magic power, the eggs the *brujos* had buried in unknown places of her sister's room were disinterred one by one, they came into her hands, and María threw them to the floor without returning them, knowing the illness did not require eggs and that the power of the mushrooms was enough. When María returned to her normal state and saw the broken egg shells she understood that it was a matter of reality not some hallucination caused by the mushrooms.

Following the miraculous treatment of her sister, María began to practice her profession of *curandera* and to gain the confidence of people. She abandoned the grub hoe and did not go back to cutting coffee. She took care of childbirths, and men who had a heat or cold in the body, she gave them back their souls which they had lost by having them scared away by evil spirits.

In her practice, María always used three types of mushrooms[N] exclusively: the one called *San Isidro*, the Little Bird (*pajarito*), and the Landslide (*desbarrancadera*). The Landslide is found on the *bagasse* (rotting chaff) of sugar cane; *San Isidro* inhabits manure, and the Little Bird prefers to sprout up in the shade of the cornfields or the plants that cover the humid forested slopes.[10]

The Death of a Son

A scene took place between María and her son Aurelio the second time Wasson took the mushrooms, which for us can illustrate the idea María has formed of the divinatory power of the mushrooms. Wasson writes:

> The Señora's behavior differed much from what we had seen the first time ... There was no dancing and virtually no percussive utterance. Only three or four other Indians were with us, and the Señora brought with her, not her daughter, but her son Aurelio, a youth in his late teens who seemed to us in some way ill or defective. He and not we were the object of her attention. All night long her

singing and her words were directed to this boy. Her performance was the dramatic expression of a mother's love for her child, a lyric to mother-love, and interpreted in this way it was profoundly moving. The tenderness in her voice as she sang and spoke, and in her gestures as she leaned over Aurelio to caress him, moved us profoundly. As strangers we should have been embarrassed, had we not seen in this *curandera* possessed of the mushrooms a symbol of eternal motherhood, rather than the anguished cry of an individual parent. But by any interpretation this untrammeled and beautiful outpouring, touched off in all likelihood by the sacred mushrooms, was behavior of a kind that few Middle American anthropologists would ever expect to see (Wasson and Wasson 1957, 301).

While interviewing María Sabina, knowing that her son had died tragically, I asked if his attitude that evening was compliant with what she foresaw about Aurelio's approaching demise. "Aurelio was sad," María explained. "That night he told me: 'Mother, I know that I am going to lose.' Don't say that, I answered him, but I knew that a tragedy was coming and I could not prevent it. After the *velada* (mushroom ceremony) that Mr. Wasson talks about, I took mushrooms with my son Aurelio and our friend named Agustín. While we were in ecstasy, a man carrying a rotted, rolled up bull hide appeared and shouted in a frightening voice: 'I have killed four men with this.' I asked our friend, 'Agustín, did you hear what he said? Did you see him?' He answered, 'Yes, I saw him. He is a son of Dolores.'" (Dolores was the name of the assassin's mother). "My son Aurelio died fifteen days later. Dolores, drunk, went by the patio running and she stabbed him with a knife." Why did she kill him? She must have had a reason. Herlinda took charge of answering me: "Aurelio was a retailer and Dolores owed him fifty *pesos*. Maybe that is why she killed him."

The Language of the Divinity

Regarding the poetry of María Sabina, that is, her shamanistic songs, we have the disk recorded by Wasson[11] in a bad moment—María was not inspired that evening—and the translation rendered by Miss Pike. This translation contains massive gaps that I tried to fill in during my second interview with María Sabina, but apart from some corrections I failed to clarify the text of the American linguist. Her inability to translate numerous passages, like the inability of the teacher Herlinda, may result more from the fact that María has created a special language, incomprehensible just the same to residents of Huautla, than to phonetic difficulties. Shamans in Asia utilize an esoteric language, and Mexican *curanderos* and priests call this *nahualtocaitl*, the idiom of the divinity. What María Sabina has created is not exactly an esoteric language, but more like a poetic language in which incessant reiterations of the psalm and litany are linked in a series of frequently obscure metaphors, in licenses and idiomatic play common in the ranks of poets and mentions of herbs and unknown animals, which compound the already considerable difficulties of the tonal Mazatec language.

María's songs keep time like the shamanic drum, which does not preclude María from occasionally resorting to the use of percussive elements. The inexact sovereign images of the ecstasy, dispersed, undulating, seem to order themselves and fill her songs with a sense of gratitude. In my third experience, I remember that coming out of her trance, after a silence, María sang again and created a melody so smooth, so rousing—each sound opened my flesh saturating it with an infinite sublimity—that when she finished, as if it had been a concert performed by the hand of a master, I shouted out, unable to contain myself: "Bravo, María!"

Heim, speaking of the power of the mushrooms, says that they heighten the silence. It is as though there is a veil of silence between the ear and the world of sounds, like an atmosphere between the light and the eye which absorbs rays of wavelengths too long or too short. The mushrooms lift that veil. Sounds acquire a peculiar vibrancy; the deaf world recovers the fullness of its orchestration and the slightest intonations of the voice, the most imperceptible nuances, are heard magnified, transported (*traspuestos*) to a plane that is no longer the usual one, as if the terrestrial atmosphere had disappeared and it was possible to contemplate without damage to our eyes the sun's corona of x-rays.

The world becomes melodious or we recover our lost ear. Idiom of the divinity. Eternal walking (*Andantes eternos*). Perfect silences the same as a melody. Tactile music, music that can be felt and seen. The hallucination of a man accused of having eaten peyote who declared before the judges of the Holy Office that he had seen: "Many little doves (*palomitas*) like lanterns (*lucernas*) and drops of water falling on my skin, as when it lightly rains."[12] Luminous doves filling space by the thousands; music transformed into rain falling upon one's naked body. Flight of doves, of fireflies, of liquid diamonds, counted as red, yellow, green, cubism, speedism (*tachismo*), forming, reforming, being born and dying, the musical motif expressed in the real images, visible, felt by each of the pores of our skin, by each bristle, each hair, each muscle, by the galvanized, electrified mass of the brain simultaneously producing and receiving that inexpressible cosmic melody.

The ecstasy is interrupted abruptly by María Sabina pronouncing repeatedly the name of her clients. In this case, my name: "Fernando, Fernando, Fernando." The teacher Herlinda intervened: it is necessary to answer her "I am here." By a superhuman effort I responded, confused: "I am here."

I now think it is cruel to startle the intoxicated one in his trance, but the call forms part of María's technique, it is a step in the ritual that possibly has as its objective an interruption of the chain of dissociations to return the patient to awareness of his personality.

At other times the calls are less personal although equally effective. There exists a deliberate intention to break up the sequence of the song, to maintain alertness in the patient or prevent him from remaining for a long time in a part of the delirium occupied by shameful recollections and frightening metamorphosis. María changes her tone, introducing a certain disorder, a complexity not anticipated, a disagreeable insistence, equivalent to passing from one extreme of ecstasy to another, to live in eternity and recover the sense of time.

PHOTOGRAPHS (no numeration) and captions (plate between pp. 240 and 241):

1. Two photographs, of mushrooms held in the fingers (above), and placed in a clay bowl (below), with the single caption: Young clowns, children who sing and dance.
2. A group of a half dozen or so participants in a velada including María Sabina, with the caption: The priestesses open the door to the mysterious.
3. Accessories of the ritual laid out on a petate *including bowls containing the mushrooms, with the caption: The mushrooms are given out in pairs, married...*

The Shamanic Songs

The power and mystery of the ecstasy impregnate the beginning of her song:

Soy una mujer que llora (I am a woman who weeps)
Soy una mujer que habla (I am a woman who speaks)
Soy una mujer que de la vida (I am a woman who gives life)

Soy una mujer que golpea (I am a woman who strikes)
Soy una mujer espíritu (I am a spirit woman)
Soy una mujer que grita (I am a woman who shouts)

Later the rhythm changes:

Soy Jesucristo (I am Jesus Christ)
Soy San Pedro (I am St. Peter)
Soy un santo (I am a male saint)
Soy una santa (I am a female saint)

Soy una mujer del aire (I am a woman of the air)
Soy una mujer de luz (I am a woman of light)
Soy una mujer pura (I am a pure woman)
Soy una mujer muñeca (I am a wrist woman)
Soy una mujer reloj (I am a clock woman)
Soy una mujer pájaro (I am a bird woman)
Soy la mujer Jesús (I am a Jesus woman)

Soy el corazón de Cristo (I am the heart of Christ)
Soy el corazón de la Virgen (I am the heart of the Virgin)
Soy el corazón de Nuestro Padre (I am the heart of Our Father)
Soy el corazón del Padre (I am the heart of the Father)

Soy una mujer que espera (I am a woman who hopes)
Soy una mujer que se esfuerza (I am a woman who makes an effort)
Soy una mujer de la victoria (I am a woman of the victory)

Soy una mujer del pensamiento (I am a thinking woman)
Soy una mujer creadora (I am a creative woman)
Soy una mujer doctora (I am a doctor woman)
Soy una mujer luna (I am a moon woman)
Soy una mujer intérprete (I am an interpreter woman)
Soy una mujer esterlla (I am a star woman)
Soy una mujer cielo (I am a sky woman) [13]

María Sabina expresses a different metamorphosis of the ecstasy, and the feeling of strength, of elevation and grandeur that the mushrooms give her. In this hallucinatory gallery of states of the spirit (*estados de ánimo*), of her own fragmented face, her suffering appears once: "I am a woman who cries." The other images, on the contrary, reflect an awareness of a sacred and mysterious power. She is the victor and of the law, of thought and life, light and air, moon and morning star, but she is also the cloud and the clock, the doctor woman, interpreter woman, and wrist woman, a woman saint and a male saint—even the sex is acknowledged in the celestial hierarchy—and something beyond saintly (*santidad*) because it is the source from which the sacred flows (*mana lo sagrado*): the heart of Christ, the same as the heart of the Father.

Neither is it possible to express that phase of the ecstasy in a more natural way. It is useless to try to reconstitute the substance of the dreams or offer an idea of the complicated and subtle architectures glimpsed under the effect of the mushrooms. Wasson, in his interpretations, follows the path opened by Huxley. Psilocybin acting upon a Western brain stirs up Western images. María is an illiterate Indian who has no relation with the world of Wasson; her thought and sensibility pertaining to the world of magic and its formal expression come from very far back, from the reiterations and parallelisms of the *Popol Vuh*, from ancient hymns, from Aztec songs, and her vigorous rhythm, the rhythm that creates the ecstasy and sacred climate, is the rhythm of the jaguars and eagles in the friezes of Xochicalco, of Tláloc and the snake at Teotihuacán, of the big-nosed heads (*las cabezas proboscidias*) of the Chak in the Temple of Kabah, the rhythm of those temples, true books in stone, where the devoted multitude could intone, through the large repeating forms (*las grandes formas repetidas*), the canticle to the divinity. "Repetition," says Paul Westheim, "here is affirmation, a method for recording the message in memory, emphasis, invocation, yearning to conjure, prayer."

On the other hand, the successive changes that María Sabina suffers undergoing are not merely the individual expression of ecstasy, but the expression of the magical environment that in spite of everything is kept alive in the Sierra Mazateca: that of metamorphosis. The *curandero* transforming into a jaguar, into a bird, a serpent, a god or demon to bring rain or cause hail, to cure diseases or smite the terrible violators of the law with punishments for their evils. The memory of the gods taking the form of a tiger, of an eagle and an owl; of the warriors who died during the battle incarnating the sun; the men being split into their Double (*desdoblándose en su Segundo*), condemned to share the destiny of their totemic animal.[0]

Ambience of masks, of changes, dissociations, of incarnations that María interprets by becoming wrist, clock, twilight, clown devil woman, holy woman clown, woman who goes like a clown.

María told me in reference to these last and dark metaphors: "I see the mushrooms like the children, like clowns. Children with violins, children with trumpets, children clown who sing and dance all around me. Children tender as sprouts, like the buds of flowers; children that suck out the bad humors, the bad blood, the dew in the morning. The bird that sucks out the illness, the good hummingbird, the wise hummingbird, the figure that cleans, the figure that is healthy."

"I sing to the sick: Here are my medicinal leaves, here are the leaves to cure. I am the lightning woman, the eagle woman, the wise herbalist (*la sabia herbolaria*). Jesus, give me your song."

Coatlicue in Reverse

The second half of María's shamanistic song begins very quickly with a litany, which as is natural, Miss Pike finds difficult to translate. The obscured names, consciously shuffled are mixed with each other in order to create confusion. They parade by galloping, striking, beating with violence until their rhythm eases and the names evoked become recognizable.

San Pablo
San Pedro
Pedro Mara
Pedro Matin
Pedro Martínez

About this game of words Miss Pike writes: "Here the interesting thing is the name of Pedro Martínez. I have the impression that it could be Martínez is being used as a last name of San Pedro, much as Christ is the last name of Jesus. Notice how she constructs it from San Pedro, to Pedro Mara, to Pedro Martínez."

María clarified the question to me by telling me that she introduced that name to honor Pedro Martínez, brother of the teacher Herlinda, in whose house the mushroom ceremony was celebrated that night. Thus also it is that Aritano García figures in the song, for Cayetano Aritano García, the town Syndic to whom Wasson turned on his visit to Huautla in 1954.[6] (Ed.—1955 according to other accounts).

After playing with San Pedro and Pedro Martínez, María, according to notes made by Miss Pike, uses for the first time the word *ven* (see):

Ven, Santo (They see, male saint)
Ven, Santa (They see, female saint)

Vengan, trece diablos (They are coming, thirteen devils)
Vengan, trece muchachas diablas (They are coming, thirteen devil girls)
Vengan, trece muchachos de la escuela por el agua (They are coming,
    thirteen boys from the school by the water)

I asked her to explain the meaning of those dark chants to me and she answered: "It was in ecstasy that Mr. Wasson felt sick and at the same time I heard hands scratching the door. Cayetano said to me: 'María, take care so that nothing happens to our friends.' I then sang:

Que el diablo no perturbe (So that the devil does not disturb)
Que vengan trece santas (So that thirteen saints come)
Que vengan trece niñas (So that thirteen girls come)
Que vengan trece niños (So that thirteen boys come)
De la escuela por el agua (From the school by the water)

The subject of purity—I am a clean woman, the bird cleanses me, the book cleanses me, it affirms repeatedly—is one of the most beautiful and insistent.

Flores que limpian mientras ando (Flowers that clean while I walk)
Agua que limpia mientras ando (Water that cleans while I walk)
Flores que limpian (Flowers that clean)
Agua que limpia (Water that cleans)

I cannot but recall, along the trip through the Sierra, two elements apart from the flowers of the Indian summer, the small yellow, pink and white flowers scattered upon the slopes of the passes like an embroidered carpet, and the tumultuous water of the canyons that formed cascades, flowing and splashing on the travelers, darkening the steps and threatening to wash out the roads.

Waters and flowers restore from the fatigue of the trip, refresh, clean. The carriers and their beasts stop to drink the water backed up in a hollow of the rocks or take a break to feel the freshness of those million particles that dance in the air attending a rainbow about the paths through the Sierra.

The sequence of the purity, cut by lack of translation, could continue in this form.

Porque no tengo saliva (Because I have no saliva)
Porque no tengo basura (Because I have no sweepings)
Porque no tengo polvo (Because I have no dust)
Porque él no tiene (Because it does not have)
Lo que está en el aire (That which is in the air)
Porque ésta es la obra de los santos (Because this is the work of the saints)

And further ahead, after saving a new gap:

No hay brujería (There is no witchcraft)
No hay lucha (There is no fight)
No hay cólera (There is no rage)
Ninguna mentira (Not at all)

These examples, the most significant, can offer a rough idea of the "poetic mode" of María Sabina, where everything seems to come by brief gusts of wind and sudden illuminations.

Two isolated verses give some idea of her solitude; of a woman who has closed herself off from the world voluntarily:

No tengo oídos (I have no ears)
No tengo pezones (I have no nipples)

She is Coatlicue in reverse. Deaf, obstructed. She does not suppress her nipples, does not mutilate them; she suppresses her nipples in the Indian imagination, that is to say, by blindly making of them an image of a monstrous class of sterility.

About her native landscape, about the mountains in which she has always lived, only these lines of an admirable poetic feeling and justice:

Tierra fría ("Cold land")
Nuestra tierra de nieblas ("Our land of fog")

Suddenly, an affirmation, like a gunshot:

Soy conocida en el cielo (I am known in the sky)
Dios me conoce (God knows me)

To already finalize an increasing sadness impregnates her song:

Todavía hay santos (Still there are saints)

and without interruption a melancholy call:

Oye, luna (It hears, the moon)
Oye, mujer-cruz-del-sur (She hears, Cross-Woman-of-the-South)
Oye, estrella de la mañana (It hears, the morning star)

Finally, "very tired, very sad:"

Ven (They see)
Cómo podremos descansar (How will we be able to rest)
Estamos fatigados (We are tired)
Aún no llega el día (The day does not yet arrive)

The Power of the Mushrooms

At the third and final interview, María Sabina, accompanied by her great-granddaughter, arrived very early. Realizing that the teacher Herlinda was not with me, she retired to the house of Doña Rosaura, next to the hotel, and removing from her knapsack some old fashioned eyeglasses and a *huipil* sat down and began calmly embroidering. It was difficult to believe that the little old lady hunched over her fabric with the eyeglasses slipping down her small nose was the powerful *curandera* María Sabina.

The teacher Herlinda, kept over at her school by a meeting, arrived at the appointment an hour after the time agreed upon. María Sabina, angry about the delay, wanted to go back to her house and it took fifteen minutes to dissuade her. I offered her a glass of rum and little by little the frown smoothed out. She had her arms crossed and her intelligent eyes awaited my question.

"When Wasson took the mushrooms for the first time in the company of his friend the photographer, you asked that they be careful not to step on a place located to the left of the altar, because the Holy Spirit would descend in that place. Does it really descend? You call it and it comes down? Can you see it?"

I expended effort so that Herlinda made her understand the meaning of my question. She reflected for a moment and then answered:

"In effect, it comes down because I invoke it. I see it, but I cannot touch it. In fact it is the power of the mushroom that makes me speak. I cannot say what that power consists of. Without the mushrooms it would be impossible for me to sing, to dance or to cure. Where would the words that leave me come from? I cannot invent them. If someone taught me to sing I would not learn. The words sprout from me when I am intoxicated, just as the mushrooms sprout in the cornfield after the first rains."

"I sang according to the people. If he is Mazatec, one of my own, I see I have more work than interests him because there is much envy in town, there are many curses. A year ago, when I gave the fungi to you, you felt bad. This in itself was because you had hired a *brujo* but since you did not accept his services and came to me, the *brujo* took revenge by causing suffering."

"It orients me in how to be for the people who take mushrooms and their needs. I must think about the freshest water, the highest trees, the most beautiful cities. Also I must fix my thought upon the one ill so that he finds something true. I must plead so that the spirits of the most remote times, since the Trinity Most Holy made the light, that they help me with their influence so that the sick understand the ideas that are necessary for their relief. I invoke the saints, the Lord of the Hills (*Dueño de los Cerros*), the Horseman of Monte Clarín, Flowing Water Maiden (*la Doncella Agua Rastrera*),[P] and then I feel like a woman saint, like a woman who knows all, like a great woman. I am outside, far from here, very far away, very high up and I do not receive anything, I do not want anything, neither does anything concern me. When I am in ecstasy, I think that many days have passed, many years, and only when morning comes and I recover my senses do I have an idea of time again."

The little grandson of María, without a doubt her favorite, thrown upon her skirt, does not take his glance away from her shining black eyes. The grandmother loses something of her gravity and smiles, passing her hand across his head. "How many relatives live with you?" "Ten. One of my daughters sews, weaves and embroiders. Another one grows maize and beans. A son is a workman and fireworks dealer. (A month ago a gunpowder mishap left him four fingers on the left hand.) Each of the three helps with the household expenses, but I put in more than all three combined. "What are you going to do?" "I prefer to walk from plans that my grandchildren are hungry. I can still work with the mushrooms now. When I am an old woman and no longer strong, what will become of us? Therefore, my greater ambition is to put a little awning around my house and to sell travelers food, beer, and small wares. I had a store, but they burned my house and now I have to begin all over again."

Archaic Techniques of Ecstasy

Can we speak of Mazatec shamanism? Does María Sabina have a relationship with the shamans of Asia? The differences, to my way of seeing, are rather in the techniques used to reach ecstasy that is the same essence of this complex spiritual phenomenon. Shamans of Central and Northern Asia arrive at ecstasy by being gradually excited and helped not so much by the drums, dances and canticles as by the same nature of the sacred elements with which they enter into contact. It is clear that achieving trance is not exclusive to mushrooms nor other narcotics, but Mircea Eliade[Q] has asked if outside the historic explanations that could account for them, these aberrant practices (decadence due to external cultural influences, hybridization, etc.) cannot be interpreted on another level. "Is it not possible to ask for example if the aberrant side of the shamanic is not due itself to the intention of the shaman to experience in concrete a mystical, but at the same time *real*, trip to the Sky, not carried by the aberrant trances ... if, in short, these behaviors are not the inevitable consequence of a frustrated desire to live, that is, to experience as in a carnal land, something that in the present human situation is not already accessible except on a spiritual plane."[15]

I put aside whether we have the right to describe as aberrations what presents itself as a simple variation of the same oriental technique. The use of the mushrooms, although mixed with Catholicism, is not only the consequence of external cultural influences, but a practice that has been conserved to weigh against Catholicism. María Sabina otherwise does not feel any desperate desire to experiment in a carnal land that today is no longer within reach of our world, because this is a bookish preoccupation or at least a mechanism of Western thought completely alien to magical thought. In the last analysis María, like Mazatec shamans before the Spanish conquest, cannot desire ardently what she performs in a normal and constant way thanks to the mushrooms. Their ecstasy, their mystical ascents, their communion with the flesh of the god and even their own metamorphosis are part of a technique, a control already achieved over certain sacred and magical elements.

In any case, what is amazing is not the variants and difference, minimal, of this phenomenon, but its unity and coherence. Around María or the $\check{c}o^4ta^4ni^4\check{c}e^4$ of the sierra—above all in isolated places like Ayautla where tourists do not visit—is centered the magico-religious life of the Mazatecs, which does not mean that this type of shaman "is the only manipulator of the sacred, nor that the religious activity is totally absorbed by it."[16]

The religious feelings of the Indians possess an amazing dynamic, and any consideration that we make concerning them would be false and distorted if we did not have what she presents. In the Sierra, María Sabina—to reference an individual case—coexists with *curanderos* of another type, snake handlers (*culebreros*), prayer men (*rezanderos*), and Catholic priests.

To the Mazatecs the Catholic religion is not sufficient for them and they need to quiet their hunger for sacred and magical elements, their insatiable voracity, with a great number of *curanderos* and *brujos* or manipulations and practices that are carried out independently of the Catholic priests and Indian *curanderos*.

On the whole, the most important thing in this religious mixture is the ecstatic experience, "considered as the religious experience par excellence."[17] Therefore the ones who dominate in the Sierra are not the *curanderos* or Catholic priests, but the ones who resort to the sacred mushrooms, being—within a variety of techniques poorly studied—the specialists "of a trance during which the soul is believed to abandon the body and undertake ascents to the sky or descents to the underworld."[18]

Another fundamental aspect of the shaman is his control over the spirits. María Sabina invokes the Lord of the Hills, the *chaneques*—gnome-like beings (*duendes*) who snatch the soul of *asustados* (Ed.—those afflicted with *susto*, or scared sick, as culturally defined)—the Virgin Mary, St. Peter, and St. Paul; she is also capable of scaring away the bad spirits—supernatural Indians or devils more or less Catholic—but her guide and strength is the same Holy Spirit. María Sabina, in the matter of divine aids, does not digress. She goes directly to what is the source of the divine, to the figure in whom the vast celestial hierarchy resides, to the Father of Christ and all creation. She invokes it and it is the Holy Spirit that descends to her cabin to remain at the left side of the altar during the hours of ecstasy. The assistants to the ceremony know that it is there, in a precise place, but they are not capable of seeing it since they lack María's power, while she sees it, speaks with it, pleads with it to make her know the fate destined for the sick and the Holy Spirit obeys conducting her on wing to the realm of the dead or lifting the veil that hides its source.

The initiation of María culminates in her intervention with elderly patients and a little later with her sister. Here also the traditional scheme of the initiation ceremony is followed in a rigorous manner: "suffering, death, and resurrection." Suffering caused by the blows of the husband, blows that injure and cause bleeding—the dismemberment of the shamanic neophyte in Siberia—or the suffering inflicted by an excessive dose of the mushrooms. Death is not only the demise of the elderly and her sister, but death as a woman and as a farmer for the purpose of undertaking a new path, and the resurrection is also dual: it is rendered in the miraculous healing of the dying and of herself, when a superior spirit shows her the Book of Wisdom and María obtains the magical powers that will make of her a great *curandera*.

The Sanctity of the Natural World

Shamans represent the sanctity of the natural world. If they maintain an influence over thousands of people, it is because they are known to have gained this eminent position by their merits. A gift, a privilege, a predestination mark the shaman and make him different from other men. The ordeals to which he submits, the incredible exploits he carries out, his physical vigor, the mastery with which he handles the diverse techniques of his practice, the courage to confront greater risks, make of him a saint and an almost extinct hero.

María Sabina should be seen within that sanctity. With communications cut off for millennia, isolated in the mountains, she is the same as the Yakuts, the Australian aborigines, or the Indians of South America, continuing to build ladders and draw mystical maps in which the entities divorced from the sky, the Earth and the subterranean world of the dead, come together more and more.

She does not become aware of what shamanistic ecstasy represents, that is to say, a nostalgia and desire to recover a state "before the fall."[19] But she ascends to the sky, speaks with the gods, maintains strict communication with the spirits, penetrates into the region of the dead—and restores the broken bridges that once connected and gave coherence to the spiritual world of man. Wise herbalist, *curandera*, singer, master of ecstasy and of the human soul, her prestige has overcome by a gift, "a force that takes hold" and it allows her to leave her body, and largely, by a life of ordeals and sufferings not at all common, by a prolonged abstinence that grants access to the mushrooms and by a high consideration of her magical powers oriented towards good and not to cause harm as is the custom of some *curanderos*.

Even the characteristics attributed to the perfect shaman agree with those María Sabina displays on all occasions, since according to the Yakut, "he should be serious, tactful, know how to convince those who surround him; and above all should never seem presumptuous, proud, violent. An inner force must be felt within him that does not offend, but that becomes aware of its power."[20]

María Sabina is indeed not serious, but serious and worthy as are almost all the Indians. In spite of the fact that the success of the mushrooms has brought about the advent of unscrupulous charlatans, of grudges and jealousies caused by commercial competition, she is not violent nor does she express bitterness toward the fakers. Far from showing pride or conceit, she dresses in a faded and even mended Mazatec *huipil*, one that shows her bare feet. Nearby, or after some minutes of treating it, she ends up prevailing (*De cerca, o después de tratarla algunos minutos, termina imponiéndose.*). A control over herself, a perfect naturalness, an awareness of her power expressed only in the wondrous depth of her eyes, combined with a calmness all figure, certainly making of her an extraordinary personality. She knows that she is famous—she keeps the pictures and articles that have been published about her—but she does not like to speak of the subject. Like all her people, she is small and thin and would be too thin if not for the visible muscles that show increasingly under her dark skin. Her hair, divided by a part, is still black, like her eyebrows, thick and abundant, a rare thing among the Indians; she has salient cheekbones, a broad and strong nose, and a large and eloquent mouth. Her life as a peasant, having maintained her family over many years, her journeys undertaken on foot and the long evenings when she exercises her profession of *curandera* singing for five or six hours, dancing and handling percussion elements, drinking *aguardiente* and smoking, do not seem to have diminished her prodigious energy.

Many Mazatecs go up to seek her in her solitary cabin, consulting her for their problems, they have faith in her treatments, and surround her with respect and consideration. María Sabina does not accord an exaggerated importance to her lofty position. Instead of surrounding herself with mystery, she is seen in the street with bulky loads or seated full of humility in the corner of the church.

Her frequentation and handling of the sacred do not prevent her from fulfilling her family duties, and in such a fashion her two existences appear united so she does not officiate in any ceremony without one of her grandchildren being present. The child sleeps curled up, like a lamb supporting its head upon its legs folded together. María Sabina caresses him from time to time and when he awakes she offers bread to him or covers him with a shawl. Tolstoy, without a doubt, would have liked to know this small old woman who speaks with God face to face, lives in a state of purity, obtains her daily bread by seeking out remedies on the mountain and curing her moral and physical sufferings, is a mystic and at the same time a woman who with great sacrifices and pains accomplishes that task so difficult—especially in the Mexican countryside—of staying ahead—just to survive—for the children, the women and the elders of her extensive family.

MYSTICAL ASCENT AND DESCENT INTO THE UNDERWORLD

Carlos Inchaústegui did not know a single word of *nanacatl* nor of its miraculous effects. He considered the subject as territory reserved to satisfy the morbid curiosity of foreign tourists, and when I communicated to him my intention to attend a ceremony, he tried to help by taking me to a fat *brujo*, with a wide and malicious face, dressed like a *mestizo*, who was the proprietor of a ramshackle shop located on the main street of Huautla.

The *brujo* took out a very dirty deck of cards, spread some out on the counter and observing them intently told us: "The cards announce there are no objections nor dangers in eating mushrooms. They can come to my store at nine o'clock tonight."

The price being agreeable, we returned to Inchaústegui's kitchen, and while we waited for our coffee, Gordon Wasson appeared in person. If Inchaústegui's ignorance about the subject was great, mine was immeasurable. Not having read a single line of Wasson neither of us suspected in the least that the man with the large head, heavy lidded eyes and ceremonious speech was the greatest connoisseur of the hallucinogenic fungi.

Wasson, perhaps feeling pity for my innocence, advised me that I ought to resort not to the charlatan I had hired as a *brujo*, but to María Sabina: "The sacred mushrooms were formerly never sold in the street, just as the communion host is not; but today they are offered everywhere and constitute a business already worth some thousands of *pesos*. It is necessary to be on guard against charlatans and pretenders. María Sabina is profoundly expert in her profession and you must bear in mind that each ceremony is an individual work of art. I recommend her for this very reason. In any case, the mushroom ceremony must be conducted in a secure and secluded place."

"In my hotel?" I asked. "No. It is an unsuitable place. It would be better to celebrate it in the house of the teacher Herlinda."

"Is there danger in taking the mushrooms?" "None. Nobody abuses the mushrooms and no one wants to repeat the experience. People only resort to them when they have problems."

Just after Wasson left I sent a message to the *brujo* canceling the ceremony. I hired María Sabina for the following day which was Sunday. I rented the house of Herlinda the teacher and, what was more, convinced my friend Inchaústegui to eat *nanacatl* in my company.

Sunday evening I went over to *la profesora* Herlinda's house, taking rainclothes and lamp, with Mrs. Beatriz Braniff, my friend Carlos Inchaústegui and his wife, and a Mazatec teacher Lucio Figueroa. Her household is situated in an upland spot of Huautla and is made up of two cabins amply separated by a patio. The darkness at night—Huautla, in spite of being a city of twenty-two thousand inhabitants, lacks electrical lighting—the presence of the trees and the situation of being surrounded by clouds and dangerous slopes, magnified the mysterious charm of that unknown place.

The ceremony was celebrated in the more spacious cabin. At the heart of the altar there stood two lithographs of San Miguel and Señor Santiago[R], a bouquet of yellow chrysanthemums (*crisantemos amarillos*), two wax candles, *pisiete* (Ed.—*Nicotiana tabacum*), and a good handful of hallucinogenic mushrooms. María Sabina and her sister were seated on the ground with three or four grandchildren, tightly bunched together and forming a small group.

PHOTOGRAPHS (no numeration) and captions (plate between pp. 256 and 257):

1. (picture of María Sabina): "I am a doctor woman, I am a moon woman."
2. "I am known in the sky, God knows me."

Beatriz, Inchaústegui and I ate our portion of mushrooms gladly and carefree, without ceasing to make jokes and knowing nothing of what was going to happen to us. Half an hour later I felt as though floating, like a feather in the air, and the first visions made me understand I had penetrated into a new world. Little gray vipers undulated rhythmically on a red background, but this vision did not have anything pleasant, it involved anxiety, a slightly distressing irrationality, an image of fever, a product of invading nausea. It was full of the poison of the mushrooms, that mineral and decomposed flavor of death. It was nothing still, nothing clearly defined. It was possible to leave, to return to the solid and coherent world, to the reason that was there, represented by the witnesses and the people who stood by and declined to eat the mushrooms. What did they matter now? Distant and blurry they symbolized another, contemptible reality, the routine one, the reality that we had voluntarily rejected.

The tiny visions of color returned. Persian tapestries, golden Chinese fabrics, oriental brocades unfurled the sumptuous monotony of their drawings in an audible, sonorous silence, the absolute silence of the high mountains and sidereal spaces.

María Sabina rendered the psalm. I recalled a song about the beam of a wine press, sung in Nahuatl: "My haunches dance although they are sunk in the water." Yes, I still had the faculty to think about that beam moved by the devil who strutted before the amazed eyes of the conquerors.

María Sabina sang. But was it truth that sang? Her voice made the tapestries and brocades undulate, gave them movement, and their drawings paraded by quickly, being made and undone in a flight that had neither principle nor purpose, going off incessantly, uniform and of perfect regularity.

It returned, always returned to the tiny world of the worms, the filiform world, the white gelatin, the putrescent swarm. Those little wriggling vipers had tiny eyes, red and green that punctured and hurt like pins, eyes that were transformed into crowns, into medallions of rubies and Hindu sapphires, into microscopic spears, brilliant thorns, all inhuman, all beyond our world, a cellular mesh composed of minerals, phosphorescent, sharp, heartrending.

It could still leave. It could leave but it did not want to depart from me. It was only a preview of the horrendous thing, of the unknown thing that approached. I wanted to speak, to register those images—why that idiotic eagerness to register it all?—to display them to posterity, to yield unto it that incomparable legacy, and was only able to say a word, a foolish word, that made me laugh foolishly.

Oh, oh, oh what hallucinatory confusion (*deslumbramiento*), what new force, what metamorphosis operated within my body. I see dawn in the bay of Havana from my room on the eighteenth floor of the old Hilton. The fog embroiders the tender blue of the coast, the sea shines pink like a silken fabric and below, in the dark well of the deep streets, the lights of the first cars slid by. Socialism had arrived, the ghost had crossed the sea and was there, invisible, between the North American sky-scrapers and advertisements for Coca-Cola. Socialism had arrived and apparently all felt equal. I had eaten mushrooms and felt equal myself, except for that irrational danger that stalked me. It should not frighten me. If I am frightened, my God, I am lost, like that morning in Acapulco when I left to look for sea stars and the undertow dragged me out to sea. To die, stupidly, far from you, María, lying neglected on the beach, your hair soaked with salt, your hot sex soaked with salt, your teeth of pearl soaked with salt, your hair humid with salt, lukewarm swamp where millions of horrendous creatures twist and proliferate. The salt water enters me in wafts, sprouting in my intestine and bursting like a wave of putrescence in my mouth. María Sabina, psalm of great shamans, architecture of light, powerful force of the spirit, always fighting against nausea and an urgent desire to urinate, but I do not have to urinate, the water has chemical substances that would denounce the yellow spot of shame and you, champions of the triple jump, world champions of the crawl, champions of hard rumps and narrow sphincters, sirens with shaved armpits anointed with deodorant soaps and mouths opened with the evenness of waves, my Lord Jesus Christ, Virgin of Guadelupe, no, I do not want to hear those words, María Sabina, she speaks in Mazatec, not saying a single word I recognize, do not send me back to reality, do not say a word that would recognize and destroy the ecstasy and return the nausea and sensation of feverish trembling.

I leave the delirium, I escape, I open my eyes. Beatriz, lying down next to me, is silent and motionless. The luminous orb of her watch shines in the halflight and its symmetry, a product of reason, calms me down. I recover time and measure it, that being a way to conquer it. I also recover space. Incháustegui is sitting down on a chair next to his wife, and her heavy legs seem to me like the columns of Chichén Itzá. It is an accomplishment for me just to sit on the *petate*. A light blinds me. A diffracted light, a light that vibrates with an unknown wavelength, an ultraviolet light, mortal, destructive of the rods of the pupil, a light that leaves cracks in the shape of a cross, rays of Jehovah blinding to the worshippers of the Golden Calf. "Put out that light," I managed to say, "it is my judge."

The Sequence of Carmen

María sings and her song opens a tunnel for me, a drain tunnel, dark, dense, oily, that carries the (phosphorescent) excrement of the virgin, of the beggar, of the archbishop, of the banker, of the saint, of the athlete, of the cancerous, and I go into that tunnel with my own excreta sliding me by those tunnels loaded with matter, with lichens, with polyps, very small green globes that explode, with pincers of crabs, mollusks, with blind tentacles that bear a green bioluminescent spot (*una lucecita verde*) at their fleshy tip, the fifty million year duration of the Pleistocene suddenly remembered, the Pleistocene fissure of the cerebrum aroused, the return to the beginning and its horror, to its cold, its nausea, its deaf combat. The palms and the drum, the palms and the song of María Sabina and the pain of having lost Carmen, drowned in the river (because she died thus and not of a malignant tumor as the Consul made believe), and it was necessary to look for her in the other world, to seek that proud girl, that strong girl of stout neck whom I dominated by sex on Sundays, removing her from the bathtub dripping with lukewarm water to make love to her on the wet sheets, while the bells of the rosary sounded below, Our Father who art in Heaven, no, no, María Sabina, do not condemn me to lose her with your Paternosters and your Hail Maries, let me see her once again for I am full of the poison of your mushrooms;

she sings in Mazatec or in Chinese, María Sabina, she claps. Oh yes, you sing. The door to the rainbow bridge tunnel now arrives and by it I ascend among the clouds and descend to the bottom of the sea. To see it there, in that space without form upholstered in plastic, of small gray tubes, incarnated diamonds, in the midst of that dense and opaque decoration, the boundary between the outer world and this new world, is to know that which is dead.

Already everything is possible. To live again in the room of the shabby hotel, like twenty years ago, in that squalid room where the only beautiful thing was her young naked body, and to speak with the dead. "Carmen I cannot live without you," I tell her. "I will come to see you Saturday evening. We will have dinner together."

The table is set. Roast chicken, a bottle of wine of the countryside, a German pie. Her shirt of embroidered silk hangs on a hook in the bathroom; on the shelf the cases of makeup, the perfume bottles, the lipstick, the eyebrow pencils. I expect her at the table. The city bustles below and the sound of the trolley cars enters through the open window. I am the same myopic and jealous young man who must register everything while she awaits the sound of the door containing the contractions of her womb. A fly begins to fly around the lamp. I have found her address book and I leaf through it in search of a name, a recent phone number. I no longer think that she will come from the other world, no. She is alive, I am young and these last twenty years, her tumor in the brain, her mysterious disappearance, have been a mere nightmare. The fly buzzes, flies around the table fearing my hand. The nausea. The persistent fly returns and I again shoo it away. It hits the lamp buzzing. Musical buzzing, deaf and rhythmic, sedative. The world has been drained. I do not hear the noise of the street. Outside it is dark, the oppressive darkness of abandonment, of heart-wrenching solitude. I have stopped the music. The fly is motionless upon a rose. I take a napkin folded double and swat at it; the vase falls over on the table cloth, spilling its water and stripping petals off the rose. The fly escapes and is going to land in my mouth. I feel its cold contact, its cold tickling sensation and I pursue it with the napkin until it escapes out the window.

I sit down again. The yellowish tubes, incarnated diamonds upholster the room and they isolate me, confine me, produce in me a terrible anguish. I am alone. I understand. I get it. That fly was her and I shall not see her again. Is there no antidote? I want to leave. I want to escape. Beatriz, give me your hand; you are the antidote. Beatriz remains quiet and her silence is a thick plastic, isolating, irrational.

To be God is to be Poisoned

In ecstasy I am not alone. I am like the children or the dogs who outdo themselves when they have an audience to entertain. I am a performer who needs a public.

Ecstasy is to be poisoned. To be God is to be poisoned. The poison is the substance of which God is composed. Give me another glass of poison. Poison equal to euphoria, equal to flight, to strength, to madness.

Labyrinth. I have the thread leading out of the labyrinth. Fishbowl. Aquarium. Am I the fish? Am I the visitor of the aquarium? I laugh at myself. Do I know why I laugh? I laugh at myself because I am a bubble, like a soap bubble, a prismatic bubble, a plastic bubble, a translucent globe, a retort, a crystalline sphere that rolls along on a toboggan of crystal, rolling, rolling with other spheres, with thousands of spheres, millions of spheres, and they fall, they fall indefinitely, they slip indefinitely into the dark space.

Rise and Fall

Inside of me there felt a powerful force. Surely—and this is a very subsequent consideration—I had eaten that type of fearsome mushroom sought by the native nobility for their banquets and celebrations and paid for at high price, since my state of being was a mixture of atrocious pride, of high regard for myself and a desire for barbaric adventures that would have been uncontrollable if my physical strengths did not betray me.

The idea of my superiority did not leave me during the first two hours of trance. It burned in flames. It was not the strength of my youth recovered, but another type of force, a new wisdom, a penetrating lucidity, a dazzling certainty of knowing all and encompassing everything, united with a sensation of elation and wild happiness that coursed through me like an electric current. God, I was God. Released within me were divine possibilities that had remained overshadowed and subdued until that moment.

Unfortunately, not even the delirium of one's own greatness, of the sudden transformation into divine being is possible to reconstruct once returned to our human condition. It ignored how one had become a superior being, with a genius that had a message, something very important to say. It spoke. It spoke standing, inspired. Behind me anthropologists, university girls, innumerable people attended, amazed at my transformation and my pointed words.

(I review the notes taken in shorthand by Mrs. Inchaústegui and find isolated phrases, imprecise, always cut.)

The space of the room was in a half-shadow. I cannot specify if there was a moon or some veiled light existed. Neither could I have been able with more light to fix my attention on the furniture, the newspaper, or the fabric of my pants. The sensation of my euphoria, of my force, of my magical exaltation was total and it completely occupied me. I did not fly. I undertook no mystical ascents, did not float in space. My feet were firmly planted on the ground. The sky was there, in that multitude who listened to me with reverence and registered even my feverish shouts and exclamations.

I did not know how long that exaltation would last and if it were continuous or interrupted by the noises, conversations or incidents that followed one another in the cabin. I only remember the laughter, the laughter that was going to undermine my sense of superiority. The laughter did terrible damage to me. It was sarcastic laughter, making fun of me and filling me with rage.

(I find my protests in the shorthand notes.) Why are they laughing? Who are they laughing at? The unheard-of lack of respect offended me and I attributed to that laughter a malicious intent. I felt misunderstood, vexed, unjustly humiliated. All present were my enemy. That ridiculous ceremony was a fraud. A trap. I had fallen into it. Something very serious against me was being prepared.

The trap. I have fallen into the trap it told me. Everything was premeditated, planned in advance. These Indians do not exist. They are staged. And María Sabina? My intelligence can do nothing against her primitive force. She is the great danger. Injustice. The magician. And the eyes. The eyes like balls, the eyes in bunches. The eyes that trespass against me, irrational, ferocious, derisive, threatening, the eyes that do not cease judging me, that do not stop scrutinizing me for a second.

The laughter was mixed with voices, with comments, with contemptuous judgments. The certainty of being examined, of being scrutinized broke through within me. The assistants were transformed into my prosecutors; that mysterious, fascinated audience became a court, in a judgment. They judged me a pretender, because beneath the mask of my apparent courage, of my resolute attitude, of my detachment, there lay a reservoir of cowardice, a hesitance, an unconquered selfishness. They knew the truth. They had lured me to this trap with deceptions. This was not a cabin but a courtroom. The world was set up to unmask me. "Here is the brave one. The revolutionary. Now he trembles. Now he is on the verge of crying. Leave him. He is not worth the trouble. He has aged. We will not leave him too quickly. He makes a fine spectacle. He makes us laugh. He should beg our pardon. He should confess his cowardice."

I tried to defend myself by insulting them. The target of my rage was my good friend Incháustegui who sat next to his wife trying to override the anguish of the trance. "Mount another farce less coarse," I yelled at him. "I am fed up with frauds. Belly dancing (*la danza del vientre*) is better and less boring. I go off to the mountains with the old *brujos* who know nothing of *LIFE* magazine or of *Paris Match*. Your excess of professionalism has lost you. These sloppily painted backdrops and fake Indians do not fool anyone. "How pedantic;" Beatriz' voice sounded behind my ears.

I heard this first recognizable word in the tense silence of the cabin as a condemnatory failure. Oh, you also betray me. I sought your understanding and you hang a label on me. It is time to stick labels on people. We tender them on the analyst's little couch, made to throw up their dreams, their fears, frustrations, and repressed nefarious impulses until all we have left is the rind, an empty shell. You disarm me and are not able to arm me (*Me desarmas y no puedes armarme*). I have heard the mill of the prayers in China. Wind. Rosaries. Litanies. The truth. What is the truth, foolish girl? Being God is the truth and you call me pedantic because I am a god. I should go to the mountain. I am expected up by Chicon Tokosho, the Lord (*dueño*) of Mazatec country.[5] Up, with the dead, with the tigers, the devils, and the happy little elf people (*los alegres du-*

endes chaneques [T]). Do you know? Flesh is the unique god of men. It makes us fall to our knees, to drag ourselves down begging, to howl at night, to give up our dignity, because given the choice between solitary dominion and love and its shameful weaknesses, we always go with the latter. But today is something else. I believed I was drinking poison and death and drank the elixir of wisdom. The wisdom was within me, overshadowed and without expression, and now it is revealed to me. I have been close to the metamorphosis. I had a premonition one night. I arose toward the Sierra Madre Occidental and at my feet there arose a red moon illuminating the silent swell of rock. Huichols accompanied me, men who fought for the land and we were together because we were brothers-in-arms in that fight. Then I rejected peyote. For me it was enough that they let me be by their side. I walked between irons of the airplanes and the tanks torn up at Girón Beach and for me it was enough. Never mind that this magic existed, this chemical substance capable of changing men into gods. How can you, Incháustegui, speak of the Indians if you do not know their deliria, if you have not sunken into ecstasy, have not descended into the underworld with them? Oh catharsis, catharsis. Empty funnel of the overloaded barrel of the unconscious, absolution for sins, baptism and communion, resurrection among the dead, their appearance in the Valley of Josafat, infinite relief. Oh changing sky, oh changing world. Mushrooms. Mushrooms. Mushrooms. Forgotten paradise of mushrooms. Where did I read about walking in a forest of giant fungi? Under the fleshy shade of the giant mushroom, under the delicate lattice of the gigantic fungi? Did you know that mushrooms walked in couples? I devoured a married fungal pair, I swallowed two fungal spouses. Oh, oh, they shouted, he chews us up on the night of our wedding. Ha ha ha. I do not know if I should regret it.

(Then someone lights an electric lamp.) No, Incháustegui, don't light that reflector. The ecstasy should be conducted in the half-light, like the great and mysterious rites of antiquity. There are too many reflectors, too many recorders and too many anthropologists studying my reactions. Why are they laughing? Surgeons don't laugh. Sodium pentathol. The light was extinguished but the sensation of being stabbed did not disappear. The flesh defends itself from the scalpel. Why do they laugh? I am not myself. Love has abandoned me and a man without love is garbage.

(I open my eyes. Next to me, the angel of death spreads his membranous wings and I send up a shout.) "Don't be frightened, master," Lucio tells me. It is the teacher Herlinda. "*Profesora* Herlinda, *profesora* Herlinda," I exclaim, begging her help. Herlinda brings in a tranquilizer: "The *brujo* is coming in the morning. He is causing suffering."

What can that *brujo* do to me? He is a phoney *brujo*. A mercenary. I am going away to the mountain. These last ten years I have lived on a mountain, the mountain of the mother of the gods. Could it be they did not know? At night, the moths with shining eyes and powdery wings of crystal flutter past. The cupolas open and in the silence sidereal clocks strike. Inside the pyramid of Cholula the Grasshopper God (*el Dios Chapulín*) laughs, the Spanish Virgin laughs, the star hunters (*los cazadores de estrellas*). I hear their laughter in the middle of the night. That is my school, Incháustegui. A hard school, you can believe that. An old woman dug with her fingernails two kilometers of tunnels inside the pyramid and discovered the face of the gods. She showed them to me by the light of a candle while the *nahuales* (Ed.: *nahual* = *nagual)* howled. Inchaústegui, you have betrayed me but you will not be able to win over me. One night, set against the volcanoes, in the Observatory of Tonantzintla, W.W. Morgan took a piece of chalk and drew two lines on the blackboard. "That is everything we know about our galaxy," he told me. Poor Morgan. He spent his life cataloguing stars, as you would file loose Indian words and he only knew two arms of the galaxy. Parlor anthropologists, measurers of skulls, collectors of potsherds, you know nothing about Mexico. I know Mexico and I know what keeps a man on the land and what prevents him from falling to pieces and being degraded. His reason and his dignity. You laugh. Thus Christ's executioners laughed at his agony. My reason. With it I escape your trap of ghosts.

Descent into the Underworld

(Here end the shorthand notes on my delirium. Mrs. Inchaústegui told me afterward she felt obligated to discontinue them because as of then—about three hours after having eaten the mushrooms—there were only offenses and meaningless phrases pronounced. It was coming from me in such a way that Inchaústegui, to be freed of my aggressions, requested of María Sabina that she have me removed outside the cabin where I remained under the rain for more than two hours. I learned all this the following day with great surprise and shame about it on my part. Nevertheless, I managed to reconstruct part of the delirium thanks to the information of my companions and, above all, to the state of extraordinary mental clarity that persisted for four days after having carried out the test. The experience came to me in its major characteristics and I could describe it by a single impulse, possessed by the fidelity and persistence of its visions, using two hours and with need to neither erase nor add a single word.)

Outside, neutralized by María Sabina, my exaltation yielded and I slowly began my descent into the underworld. From being a god, I went on to become a trembling elder, condemned forever to the irremediable decay of old age, to its weakness, to the humiliation that I knew with others feeling sorry for witnessing my total annihilation.

Of course, I did not know where it found me, neither did I think about the mushrooms, nor associate them with my present state. It had simply aged. I was an elder, and not even an elder, which after all would be tolerable, but rather a nuisance without dignity, invaded by childish fears, who trembled, shaken by an anguish and an intolerable cold.

Lucio, sent by Inchaústegui, took me by the arm, obliging me to enter the bordering cabin. While wanting to let go of the door, I tripped and was on the verge of falling exactly as a decrepit old one.

In that slow final route through the past it was necessary to reconsider love in a new light. Of course, the love, as all the subjects the delirium offered me, was not exactly a loving but erotic theme, because far from presenting me this time with real and concrete women, women who had played a part in my life, it caused me to suffer impure fevers of adolescence, where the loving impulse remains reduced to explicit visions, to pornographic visions of women who arouse us when we are adolescents. In a word that martyrdom of sex without food, reduced to its solitude that youth in our cities suffer where flesh is considered as a shameful sin, the flesh or rather its fever, its irritability, its sad anxiety or complacency; love, the eternity of love reduced to a small obscene rubbing.

And this adolescent eroticism, already forgotten, came to add itself to the eroticism of the elder who looks only for rest after having satisfied that inopportune need, exercised in secret because it is the residue of a still latent youthful force under the generalized decay of the withering and repugnant body.

I sought rest. My legs shook and could barely keep me standing. The intoxication had not left. Yes, inexplicably I was an old one. But where does old age lie? It lies in arriving at it, without notice, without signs, in a way so rare that life appears cut in two halves and we become aware of having crossed over without being able to return to the other half and recover it.

Possibly it was thus. I conserve a very vague impression of this new delirium. A succession of shameful acts, of cruelties against defenseless people, the uselessness of my reason, the pretenses, all that eating away at me on the inside, caused me to suffer frightfully.

External stimuli continued building within me with disproportionate violence. I was sunken in the delirium, to the bottom of the delirium, when the door opened and there appeared Beatriz. "Perhaps you do not know who you are?" she asked me. I took off my glasses and threw them to the floor breaking them to pieces. "I see," I exclaimed. "I do not need glasses."

It recovered the passing greatness. The greatness is not in the present, in our present misery, but in the past. Something we had of greatness. Something we had of happiness. When? Perhaps when he was young. Perhaps when he was loved and belonged. I should trust in my reason. Which reason? I have lost it. I should trust in love too, in human solidarity, but I am single. One is always the crucified. Always the one condemned to lie in hell. We are hell. Those eyes, those thousands of eyes, eyes without a face, that watch us coldly, irrationally, as the unique and irrational eye of the Most Holy Trinity looks at us.

Later everything is erased. I must have sunken into another delirium. I descended to the kingdom of the dead and would never leave the dark region. It would repeat, it was going to drag me into the subsoil, the basements, the latrines. At times I thought about my humiliation and my defeat. Sometimes I thought about my condemnation, in my eternal continuance within that world reportedly populated by horrendous creatures.

Then someone took me to the cabin where they all slept. They laid me down next to Beatriz and covered us both with warm blankets. The acrid scent of the mushrooms filled me. I was still dead, among the dead, and could not return to life, could not resuscitate.

Just then, the golden light of morning entered by the door and was reflected in Beatriz' face. I saw her blue eyes full of tears, her white skin soaked in tears, her strong and curly hair, her thin lips that parted disclosing white and shining teeth, and a feeling of tenderness and mercy invaded me.

Opening the Channel

But what remains of this excoriation, this barbaric purge, this catharsis that the soul has squeezed out of us making us vomit the poisons swallowed throughout life? Is it worth the trouble of being contemplated in an open channel, with one's guts showing his own excrements? To witness the slaughter of one's own being is of course an atrocious spectacle. One recoils from this abrasion and at first even tries to blame others, the witnesses of our humiliation, even murdering them if it were possible, because inside of us lives the hypocritical convention that we are saved if we have not made a record of our intimate degradation. Then, after the delirium, we fall into a state of unbearable depression when confronting the certainty that we are not as alive as we believed before the ordeal, that inside of us, inside that temple of the Holy Spirit, matters undergoing decomposition proliferate, by which too many moral heart attacks have killed extensive regions of the heart and we carry not one but many corpses to the hills.

Knowing that we carry a corpse signifies not the healing, but the principle of the healing. When seeing death it raises as much for me. I noted that all decay had originated, in large part, the cowardice, the fear of losing the woman I loved, the fear of staying and getting a nose broken from a punch, the fear of losing the esteem of the other.

The instinct of conservation, of cruelty, and of greed has been covered by many masks, many makeup jobs to disguise and hide from us its ugliness, but the sacred mushrooms cause those masks to fall away and reveal the instincts without makeup or masks.

"Knowledge of the abysses," the eddies, the landslides (*desbarrancaderos*) of the mountain of the *brujos*. If psilocybin brings with itself the entourage of unexpected reminiscences, mine were thrown upon me brusquely, mixing the bitter with the sweet, the first with the last.

Nevertheless, nobody should request of the mushrooms a miracle for himself, no one should go to the magical mountain expecting salvation. The answers from mescaline, from psilocybin or from the more powerful LSD will always be personal and nontransferable. Each one atones for his past and each one finds the door to escape from his jail. Otherwise, no knowledge is given to us if the will to know does not exist within us, no drug saves us if we do not want to be saved.

A New Trip Around Myself

Duirng my second visit to the Sierra, in summer of 1962, María Sabina did not want to come down to Huautla nor did she agree that the mushroom ceremony be held in the teacher Herlinda's house. She demanded four hundred *pe-*

sos, bread, cigars, and a bottle of *aguardiente*, but at the same time she offered a ceremony conducted up in her high cabin—at the summit of one of the mountains that dominate Huautla, where two nieces and a granddaughter would participate as assistants and singers.

The proposition, formulated through intermediaries, had its pros and cons. We considered the risk—subsequently confirmed—we would lack the necessary order and retreat, and in addition there was a disadvantage for us in being trapped in the cabin for more time than we wished. On the other hand, there existed the appeal of carrying out the ceremony under the private direction of María Sabina, in the stealth and authenticity of the Mazatec world. I decided to run with the adventure in all of its risks and at seven o'clock at night, mounted on mules and horses, we began the march with my sister, Mrs. Zumalacáreggui, a friend of hers, the astronomer Enrique Chavira and the teacher Herlinda in charge of taking care of the women.

PHOTOGRAPHS (no numeration) and captions (plate between pp. 272 and 273):

1. (María Sabina in prayer pose, portrait closeup): "We are tired, the day does not yet arrive..."
2. (Portrait closeup of young girl): Tomorrow, she also...

Although I already knew the chemical effects of psilocybin, the previous delirium weighed obsessively upon me and I did not have the least desire to revisit the wells of the unconscious nor to attend, like a forced witness, the parade of my reminiscences. I continued to think that taking the mushrooms was equivalent to buying a ticket and to travel around oneself–as one might buy a ticket to travel around the world—to travel on a long trip through that same region where there are no guides, neither maps, nor possible itineraries. I was afraid of that spectral flight through the thin skies and the abundant hells that integrate my past, and the irrational anguish of the trance, and at the same time I was determined to undergo the ordeal since the experience of the previous year was, with its pains and derangements, a new experience that helped me to know myself and by which I felt spiritually enriched.

A medical friend, a doctor, Raul Fournier, had proposed to administer to me a dose of psilocybin at the School of Medicine, but it did not interest me to swallow tablets in a civilized place while listening to a Beethoven concerto. The mushrooms united with the landscape of fog, with its own magical rites and the religious atmosphere of those ancient celebrations, interested me. The first time I suffered too much because I took an overdose of the mushrooms—too much for my nature—and I did not submit cooperatively to the technique of María Sabina. Ecstasy has a technique and if it does not one must invent it. Against the irrational anguish that causes the dissociation of the personality and excessive number of reminiscences, against that bistoury (Ed.—a small surgical scalpel) that scars us with the discovery of the solitude, the frustration, the animal instincts of man—Sahagún's pathological picture—perhaps there was no antidote other than to take the hand of a woman and descend with her to the circles of our inner hell, feeling protected by her tenderness, love already the uniquely positive thing in the midst of the negativity surrounding us, the unique thing that can save us from eternal condemnation.

One should trust in the shaman who directs the ecstasy, for that is the importance of his election, and if it is not possible to create an atmosphere of intimacy and retreat—María destroys part of this by being accompanied by her relations—having to ignore noises and disturbances that are alien to the ceremony is the inevitable result. María's songs and clapping—her incitements and prostrations, her joy and sorrow, the abrupt breaks and imperative calls—conduct the trance and interfere naturally with the voices, the laughter, and the snores within the brain altered by psilocybin; the one who has eaten the mushrooms hears those noises and interprets them according to the logic of his delirium, and the interferences are often responsible for the rage, the ridicule or humiliation that weakens the trance.

Things therefore appeared very different from how they were presented in 1961. By my side I had the tenderness of the two women who accompanied me, and their desire to help me as guides in the descent to the underworld was of an essential importance for me.

The moon was waning, and would take two hours before appearing. The Milky Way seemed to ascend impetuously in the deep nocturnal blue, the sky rising with it. Huautla remained down below. Its faint lights, small spots static and yellow, hitting the plants deaf to the acute barking of dogs that banished the silence of the heights, and the clouds of Sagittarius, the scintillating star clouds of the center of the galaxy, were imbued with intense life in the high clear sky. At the mountain tops great masses of clouds appeared crashing around their slopes, but that fascinating spectacle, that grandeur made of infinite superimposed grandeur, also appeared diminished—run over would be the right expression—before the clouds of Sagittarius, before that sidereal high tide, of stars beating, advancing, going back down—tides, undertows—from the abysses of dusty darkness.

From time to time cabins arose and dogs barked. I then recalled the test that awaited me and its unforeseeable consequences did not cease distressing me.

Sinking back and forth between these considerations and contemplation of the Milky Way, we arrived at the house of María Sabina. Her house is a simple wooden shack with tile walls arising at the edge of the road that leads to the banks of Santo Domingo. It is divided into two parts; the rear served as a bedroom for six or seven grandchildren, and the front part—separated by a division of boards—served also as the alcove and room to hold the ceremony. On the altar, lacking offerings, rested the mushrooms on a banana leaf along with a ceramic incense burner, candles, *pisiete*, and a flowering branch.

María recognized me in the act. She came forward with her light step and taking my hand spoke to the teacher Herlinda in Mazatec without ceasing to look at me. "María Sabina says that you must be calm," said Herlinda. "This time there will be no interference from *brujos*. Everything will be different."

The cabin was full of people. Children appeared naked in the middle of the curtain covering the doorway. One of the nieces, a youngster thin and pale, with big sweet eyes, who was taking care of a small boy, was revealed to be a good companion of María Sabina. She had an impassioned voice and her youth, combined with her recent motherhood, presented a distinct contrast with the hoarse voice and austere old age of her teacher. The other niece had an angular face and hard, shining eyes. Already neither wore a *huipil*. Although all three knew the shamanic canticles from having listened to them repeatedly, when María became quiet giving them an opportunity to participate, they sang Mexican songs or prayers that the devout intone in church.

Also present were two men of the family, her sister María Ana, and perhaps four or five other relatives who each entered and left within a short while. The astronomer Chavira sat upon a beam leaning against the wall of the shack, and the three of us occupied the center sitting on *petates*, accompanied by the teacher Herlinda.

María incensed the mushrooms and offered six pairs to each of us. We ate them slowly with small chocolate bars and waited. Everyone spoke aloud; the children ran and screamed unrestrained. The *curandera*, composed within herself, took *aguardiente* and smoked without a break. She also awaited the miracle. Within fifteen minutes I experienced an intense cold. The covers and available *sarapes* they threw over me were of no use. I trembled without being able to contain myself as though with an attack of fever. María approached me with *pisiete*. She applied the mixture to me at the joints of my arms and legs shaking with the spasms. Her serious face, furrowed with wrinkles, was close to me. Mazatec psalmstress. The teacher Herlinda told me that the cold would not take long to go away. I should have confidence and dispel every worry.

Then María Sabina returned to the altar and sat down on the ground. Apart from the violent chills that continued to shake me I felt neither queasiness nor malaise. Lying down face up I saw the thin parallel beams of the ceiling illuminated by the dim light of the candles. Suddenly the beams changed. Along their edges they showed a double row of rubies faded but sufficiently visible to transform the cabin into a palace from the Arabian Nights (*Las mil y una noches*). The chills vanished. "That" was present; the magical touch unfolded its irrational magnificence.

The normal voice of the teacher Herlinda: "Question María if you see something." "Yes, I am beginning to have hallucinations." The candles of the altar were extinguished and the priestesses sang. The new adventure began.

Approaching Ecstasy

I discovered within the same thing, hallucinatory confusion (*deslumbramiento*). The body of the man no longer contained alien sensations, no longer ears, eyes, skin, no longer coarse senses to grasp the external world of light and sounds, of heat and cold, but a new body, an instrument that joins with the universal orchestration of things living and dead, a pupil opened to other invisible wavelengths, an insane eye that creates their forms their colors, an aesthetic for which it was hungry; an eye capable of reproducing a drawing, of recomposing at its whim the lights upon which it is woven, that complex and delicate fabric in which blue and green predominate, to take down a luminous flag, to open a window where the yellow and vibrant suns of Van Gogh blaze.

My head. My great boiling head, my great head that hangs from a tree, like a golden beehive. My head, a globe that floats in the galaxy, at the dark and mysterious edges of the clouds of Sagittarius, my crystal head, my head of soft bones that break apart smoothly, sweetly, naturally, into ears, buildings, horns and cartilaginous trumpets to better produce and better receive this music, this music created by the mushrooms, by these singing children, by these violin children, by these horn children. By these blonde children who elevate me, transport me, rock me and lull me, causing me to sigh with joy. The Cantory (*La cantoria*), the Cantory, the Cantory, piece of marble broken into fragments, a liquid bas-relief to collide with the yellow suns and return me to reality, to the dark cabin, to the ground, to the mat, the sobs, the foolish laughter, to the moans of the real-world children, to the snoring of men sleeping, to the song of the roosters, and the moonlight that enters by cracks in the door.

My real head, my great thinking head—thinking deliciously—leaves the waters of dreams, becomes aware of the world and awaits, as during an intermission, the continuation of a concert without a program, which nobody knows when it began and in what form it will end.

To one side my sister laughs. She tells me she is laughing to the point of crying. At another side her friend sobs and speaks of wanting to go back to her house. I need make no effort to understand them. I comprehend them very well. I also understand what is happening in María Sabina's cabin. A boy vomits next to the altar. A singer expectorates noisily. The women speak in Mazatec. Nothing bothers me. No noise is able to disturb my Buddha-like placidity. My dream receives all its magical significance while delimiting it, limiting these manifestations of my world, a world that I recover pleasantly and lose again listening to the start of a new canticle:

> *Chjon nga santa na so* (*Chjon nga* I am a female saint)
> *Chjon nga santo na so* (*Chjon nga* I am a male saint)

Again I weigh anchor and my boat puts out to sea, this boat that is myself, this aerial and submarine boat that floats in the blue space, rowing without oars, without passengers, without a helmsman of this emptiness, the divinity. I myself the divinity, without beaches, without shores, myself empty, myself eternity, myself the universe, before the formation of the cosmic dust, the cosmic gas, the cosmic galaxies.

The Metamorphosis

And in this emptiness, suddenly the Miracle; the form without form; the Sign of the Cross (*el Signo*) that cannot remember, the key to this great mystery, the waves, or rather, the lines of the waves, the profile of the waves, the waves of the waves, without soundings, unfathomable, coming down from, from above—which way is up and down?—being borne one from the other, strangely connected, quickly fleeing, being made and recomposed, becoming fluid, transforming themselves into music, this indecipherable music, this orchestra never heard, impossible to retain, impossible to hear on Earth.

Music, *maestro*. Oh, Doctor Faustus, I have become a brilliant musician. If I only had the power to take a pencil and capture these notes, I would gain immortality. But what is immortality? Is that painful and unenviable obsession going to destroy the only ecstasy I have had in my life?

(The ecstasy is given to us in exchange for not transmitting it, in exchange for not bequeathing it, in exchange for delivering us to it without subsequent utilitarian intentions.)

Who am I? Of course, I am not the one who I was "before." That continuity, that coherence of being had broken itself into a thousand pieces. I am and I am not. I am here and I am not. I am actor and witness. I have gone and I am absent from everything, far away from all, and nevertheless I am present here and I attend to my own metamorphosis astonished by it.

It can melt me like a snowman, can make me out of glass without fear of breaking me—I laugh much at myself thinking about Vidríera the lawyer (*el licenciado Vidríera*)—I can transform myself into a plant—I feel large green leaves are being born impetuously from my left foot—I am able to swim or to fly in the changed air, an extremely smooth air, by a lukewarm substance, comfortable, benign and forming a part of her diluting me in her, in her indescribable beatitude in which appears, slowly forming, the Sign of the Cross.

The sign, the key image, the revelation. Of what? Of what mysterious thing? Of what coded language, of what new world? I could not remember in the least of what the formidable sign consisted. At times it hurts me in some part of the brain like a brief flash of lightning and when I try to fix it, it has been dispelled and remains only as a dazzling and delicate sensation, an architecture—similar to a mobile by Calder, but infinitely less heavy—an architecture of great refinement, made of rubies, rubies that burn like fires of popular devices when extinguished and shine on in the night in its abstract frame of reeds.

María Sabina sings and claps her hands. She claps smoothly. It is a dull and rhythmic music that now takes a precise form: that of a growing seashell. A white shell, unfolding, opening itself like a white flower, like the wings of a dove being quietly shaken, without taking flight, without rising, fixed in the darkness of the cabin by an inexplicable phenomenon.

The canticle becomes more powerful and the unfoldings of the shell are erased. They are entirely erased. The canticle tries to construct a new Sign of the Cross, a sign that can be a source of blue glass or simply the image of a source, the image of a childlike dream, a luminous arabesque, a baroque arabesque, a golden altar, an arching column, an altar without saints, an altar of columns singing out, with corkscrews, with golden vines, with corkscrews that begin to move and rotate, like a shuttlecock, while the organ sounds, the great organ of the Chinese chorus with mouths like the pipes that correspond to thousands of mouths with the rigidity of metal—but they are singing, they are flooding me with this deluge of songs, thousands of angular mouths, mouths of the audient, of inquisitors, of Spanish friars, dry mouths, hard mouths that sing their eternal psalm: Holy Spirit, Holy, Holy, Holy, Holy Spirit, to be dispelled—oh Spanish colony, oh sin, oh to live in mortal sin—transforming themselves into other mouths, mouths of the frescoes of Zacuala, the drawings of the mouths of the palaces of Teotihuacan, of the fleshy Olmec mouths with closed lips, transparent, allowing the even teeth of the Indians to show, the mouth feathered with blue, the mouth trimmed with triangles, of teeth of the Sierra, the matrix-mouth (*la boca-matriz*), the vagina-mouth, the mouth like a conch in the center of the fresco, the mouth that creates their blue peaks, the teeth of the Sierra; the mouth that evokes the black eyes, the black eyes trimmed with green eyelids, the beak of the bird of Tláloc, the body of Tláloc, the body of Tláloc made of stars, of seeds, of seashells, of feathers, of arabesques, of border motifs (*grecas*), the hands of Tláloc melting into large drops, that are unmade, that break up into thousands of hands, into thousands of soft arabesques that flow, disappear and are born of some others; always to leave and then return to begin and flee without any goal, without end, without destination, always this expansion—like that of the universe—to roll in curved spaces, these sidereal flights, these trips on magic carpets of oriental magi, these visions surprised by a hole in the clouds, by the breaks in the waves, the interruptions of the waves. The Sign of the Cross. The Sign of happiness, of serenity, or rest within movement, of motionlessness in the expansion and in the flight. Scintillations. Twinklings. The music—the voices—sounding here, there, sprouting from all corners of the dark cabin, from the ceiling that has been made to sing, from the *petate* that has been made to sing, from my head that has been made to sing. And there I go, kaleidoscope, crystals, rhombuses, cubes, triangles, pyramids, geometric forms in this new expansion that again commences receiving a new rhythm. Then everything melts, contracts, is grouped for the first time, a concrete vision: the pyramid of Teotihuacán, its apex truncated and perfect, its mass golden, surrounded by crestings, by stoops, temples, platforms, with friezes that melt, cool off and disappear in aesthetic delight, in a sensation of intellectual placidity united with an increasing physical well-being.

At midnight "that" disappeared just as it had appeared. The rubies that had transformed the cabin into a real camera were extinguished, the ceiling recovered its squalid appearance—the candles on the altar had been lit—and I felt the return of the solid and coherent world that has been familiar to me for forty years. I sat on the *petate* with ease. My legs had recovered their normal flexibility and strength. Nevertheless, a sensation predominated within me that was a mixture of serenity, rare confidence and insight.

The Sign

I felt that a spiritual presence had cleared me and that this presence had not vanished and that it persisted within me like a benign dream when we have still not awakened completely. My state was not of one of euphoria, but of extraordinary placidity. I had wanted to communicate it to the others, but the few words I could say bore no relation to my state of spirit. I noted with clarity the abyss that existed between the loftiness and purity of my thoughts and the clumsiness of my stammering discourse.

Outside the ecstasy I continued in a certain way to be possessed by a divine spirit, which compelled me not to action—to speak or to write—but to contemplate my inner joy. I regretted having left the trance so quickly and wished to fall into it again, but this was no longer possible. The women laughed, cleared their throats, coughed, the sick children, crying, looked for the laps of their mothers the priestesses. I exited the cabin. The waning moon was at the zenith. I slowly traveled along the path that leads to the top of the high mountain. The silver-plated clouds, round and dense, covered the masses of the dark *cordilleras* and the canopies of the trees arose from the abysses shading the edges of the path. María Sabina's cabin was at my feet. A surprising sidereal peace reigned, the majestic silence of the high altitudes, a peace and a silence trespassed upon by a fear that I was not able to dominate at that moment.

The crossroads, adorned with flowers, the gods and the spirit lords (*los espiritos dueños*) of the hills, of the springs and cliffs, the howls of coyotes, the invisible presence of the *nahuales* and the dead, was an atmosphere that surrounded the cabin, the sacred atmosphere of fears and deliriums in which the ceremony of the hallucinogenic mushroom grew like an orchid. I had the sensation of having participated in the communion of *nanacatl,* in that rite to which only the pure have access, those who have been cleansed of their sins in order to receive in their body the flesh of the old Mazatec gods.

It was not merely the chemical effect of the psilocybin, its well studied alteration, but something else of a different nature. The ecstasy of the mushrooms transcended my knowledge, my western logic, and it made me think that the communion celebrated at the summit of the solitary mountains, directed by María Sabina, "the one who knows," inside a miserable cabin, for me approached the spirit of the Mexican priests, not only the spirit of María Sabina, but also the spirit of the magi, of the diviners, of *curanderos*, of Toltec, Zapotec, Mixtec, Mexican shamans, of those nights when near the statues of their gods, inhaling the *copal* and the perfume of their flowers, they ate the sacred mushrooms and sank into its deliriums and spoke with the gods and the dead.

Tláloc was there next to me, and Nindó Tokosho and Coatlicue. In my brain, like a wound, persisted the Sign, the abstract image of the Sign, resplendent of another world, that refined structure I was unable to reconstruct or evoke, but whose magical effects, like that of music, still impregnated all the cells of my body.

It had discovered in me—there is no other form of knowledge—the ecstasy maintained in secret over the span of the centuries; the idols hidden behind the Christian altars; the umbilical cord that the conquerors believed themselves to have severed and through which the Indians maintained a relationship with their destroyed world, with the source of the colors, the drawings, the ancient forms.

The Indians have given us, not their paradise, but their knowledge. The incredible possibilities of man, of his body and his spirit, the faculty of breaking the chains that bind us, of obliterating our prison, of dissociating into several, infinite personalities that integrate our consciousness, of the collective, the rearward, the lost keys to the millennia, of the complex present, with its anguish, its insecurity and its strength and the personalities of tomorrow, ungerminated seeds of the future, in short the revelation of what man could be if he can manage to conquer the monsters created in his own imagination.

*   *   *

NOTES (Notes designated with numbers are from the original by Benitéz; letters of the alphabet designate those added after translation into English.).

Benitéz' notes

1. Prologue by Roger Heim in *Les Champignons hallucinogènes du Mexique*, Roger Heim and R. Gordon Wasson. Editions du Muséum National D'Histoire Naturelle. Paris, 1958.
2. (text of Eunice Pike's letter):
Huautla de Jiménez
Oaxaca, México
March 9, 1953

Dear Mr. Wasson:
    I'm glad to tell you whatever I can about the Mazatec mushroom. Some day I may write up my observations for publication, but in the meantime you may make what use of it you can.
    Mazatecs seldom talk about the mushroom to outsiders, but belief in it is widespread. A twenty-year old boy told me: "I know that outsiders don't use the mushroom, but Jesus gave it to us because we are poor people and can't afford a doctor and expensive medicine."
    Sometimes they refer to it as 'the blood of Christ' because supposedly it grows only where a drop of Christ's blood has fallen. They say that the land in this region is 'living' because it will produce the mushroom, whereas the hot, dry country where the mushroom will not grow is called 'dead.'
    They say that it helps 'good people' but if someone who is bad eats it, 'it kills him or makes him crazy.' When they speak of 'badness' they mean 'ceremonially unclean.' (A murderer if he is ceremonially clean can eat the mushroom with no ill effects.) A person is considered safe if he refrains from intercourse five days before and after eating the mushroom. A shoemaker in our part of town went crazy about five years ago. The neighbors say it was because he ate the mushroom and then had intercourse with his wife.
    When a family decides to make use of the mushroom they tell their friends to bring them any they see, but they ask only those whom they can trust to refrain from intercourse at that time, for if the person who gathers the mushroom has had intercourse, it will make the person who eats it crazy.
    Usually it is not the sick person nor his family who eat the mushroom. They pay a 'wiseman' to eat it and to tell them what the mushroom says. (He

does so with a loud rhythmic chant.) The wiseman always eats the mushroom at night, because it 'prefers to work unseen.'

Usually he eats it about nine o'clock and it starts talking about a half an hour or an hour later. The Mazatecs speak of the mushroom as though it had a personality. They never say, 'The wiseman said the mushroom said.' They always quote the mushroom direct.

The wiseman always eats the mushroom raw; 'If anyone cooks or burns the mushroom it will give them bad sores.' There is no specified number of how many he should eat, some wisemen eat more than others, usually they eat four or five. If he eats a lot, it 'wants to kill him.' At such a time the wiseman falls down unconscious, and comes to little by little as the other people in the room 'pray for him.' This may also happen 'if has had intercourse too recently.'

When all goes well, the wiseman sees visions and the mushroom talks about two or three hours. 'It is Jesus Christ himself who talks to us!' The mushroom tells them what made the person sick. He may say the person was bewitched; if so, he tells who did it, why, and ho. He may say the person has 'fear sickness.' He may say it is a sickness curable by medicine and suggest that the person call the doctor.

More important, he will tell whether or not the person is going to live or die. If he says he will live, then 'he gets better even though he has been very sick.' If he says he will die, then the family start making arrangements for the funeral and he tells who should inherit his property. (One of my informants admitted, however, that the mushroom occasionally makes mistakes.)

One of the 'proofs' that it is 'Jesus Christ himself' who talks to them is that anyone who eats the mushrooms sees visions. Everyone we have asked suggests that they are seeing into heaven itself. They don't insist on the point, and as an alternative they suggest that they are seeing moving pictures of the USA. Most of them agree that the wisemen frequently see the ocean and for these mountain people that is exciting.

I have asked what the wiseman looks like while under the influence of the mushroom. They say that he is not sleeping, he is sitting up, with his eyes open, 'awake.' They say he does not drink liquor at the time, but that he may in the morning. Some of them go right out to work the next day, but some stay home and sleep 'because they have been awake all night.'

Although we have never been present when the mushroom was eaten, we have observed the influence it has on the people. One of our neighbors had tuberculosis and was coming to us for medical help. Then one night they called in the wiseman to eat the mushroom in his behalf. It told them that he would die.

The next day the patient no longer had any interest in our medicines; instead he began to affairs in order for death. He quit eating solid food, restricting himself to corn gruel. About two weeks later he refused even the gruel, accepting only an occasional sip of water. A few days later even water was rejected.

In less than a month after he had consulted the mushroom he was dead. Another neighboring family had a series of sicknesses. They consulted the mushroom for their twenty-two year old son. The mushroom said he would get better, and he did. When their eighteen-year old daughter became ill, they consulted the mushroom. It said she would get better and she did.

Then the ten year old daughter became ill. The mushroom said that this one would die. The family were amazed because her illness had not seemed serious. Of course they were grief stricken, but the mushroom said, 'Don't be concerned, I'll take her soul to be with me.' So, following her mother's instructions the little girl prayed to the thing talking to her, 'If you don't want to cure me, take my soul.' A day or two later she was dead.

Not all the Mazatecs believe that the mushroom's messages are from Jesus Christ. Those who speak Spanish and have had contact with the outside world are apt to declare 'It's just a lot of lies!' Most monolinguals, however, will either declare that it is Jesus Christ who speaks to them, or they will ask a little doubting, 'What do you say, is it true that it is the blood of Jesus?'

I regret the survival of the use of the mushroom, for we know of no case in which it has had beneficial results. I wish they'd consult the Bible when they seek out Christ's wishes, and not be deceived by a 'wiseman' and the mushrooms ...

Wishing you success in your research, sincerely,

(signed) Eunice V. Pike (Wasson and Wasson 1957, 242-245)

3. In the Mazatec language, ^{1} is the highest sound and ^{4} the lowest.

4. All of the Wasson text is taken from the work cited.

5. R. Gordon Wasson, *The Hallucinogenic Mushrooms of Mexico and Psilocybin: A Bibliography*. Botanical Museum Leaflets. Ed. Harvard University, Cambridge, Massachusetts, 1963.

6. Gonzalo Aguirre Beltrán, *Medicina y magia, El proceso de aculturación en la estructura colonial*. Ed. Instituto Nacional Indigenista, México, 1963

7. Aguirre Beltrán, *Op. cit.*

8. *Ibid.*

9. Wife of the Syndic Cayetano García.

10. The species considered sacred in Huautla are described by Wasson as follows:

A. *Psilocybe mexicana* Heim. A small mushroom, dark in color; it grows solitary in cornfields or pastures. It is highly esteemed by the *curandero* who eats 15 or 20 pairs. The Mazatecs of Huautla, when they speak in Spanish, call it *angelito*. In Mazatec it is specifically referred to as 'nti^1 ni^3 se^{3-4}, a name whose first element means "*pájaro*" (bird).

B. *Stropharia cubensis* Earle. An attractive mushroom with a cream colored cap that grows in manure. Among the Mazatec it is the least esteemed of the sacred mushrooms. In Spanish they call it *honguillo de San Isidro Labrador*. In Mazatec it is called 'nti^1 $ši^3$ tho^3 and 'e^4 le^4 nta^4 ha^4.

C. *Psilocybe caerulescens* Murrill var. *mazatecorum* Heim. This species grows abundantly in sugar cane *bagasse*, singly or in great clusters. Its name in Mazatec is 'nti^1 $ki^3šo^1$, the *desbarrancadera* (landslide) mushroom. María Sabina explains the name $ki^3šo^1$, landslide, as follows: "Before sugar cane was here, it was hunted in places where the ground had collapsed." There are two classes of mushrooms called ni^4se^{3-4} (Ed. note: I have corrected an apparent typographic error in the original text by which this is rendered as ni^4se^{4-4}), the hallucinogenic 'nti^1 $ši^3$ tho^3 ni^4se^{3-4} and the $thai^3$ si^4se^{3-4} (sic; note si^4 instead of ni^4) which is *Schizophyllum commune*, sold in large quantities in the market during the rainy season as a seasoning for soups. These two types of mushroom, the one sacred and the other merely edible, are small in size compared with the other species and, according to the Indians, it is the small size of the one that inspired its name.

D. *Conocybe siliginoides* Heim. This species has disappeared from the vicinity of Huautla as a result of the present deforestation. Our Indian friends brought us five specimens from San José Tenango, a region located six hours from Huautla. It grows on the wood of dead trees, and the Mazatec call it ya^1 'nte^2.

11. Wasson, R. Gordon, and Valentina P. Wasson. *Mushroom Ceremony of the Mazatec Indians of Mexico*, recording by R. Gordon Wasson with translation and commentary by Eunice V. Pike and Sarah C. Gudschinsky. Folkways Record and Service Corporation, produced in 1956.

12. Aguirre Beltrán, *Op. cit.*

13. Translated into English literally, but my translation and even the English translation do not give the least idea of the powerful rhythm that María Sabina uses. To take the simplest example: *Soy un santa, soy un santo, soy una mujer espíritu.* Miss Pike translates this into English as: I am a male saint, I am a female saint, I am a spirit woman. And María Sabina: *Chjon nca santo-nasto, Chjon nca santa-nasto, Chjon spiritu-nia tso.*

14. María Sabina incorporates Spanish words or words that she invents. She refers to the Virgin as the shepherdess, as María dozen (*María docena*), María fable (*María conseja*), or Saint María Various (*María Santo Vario*). She speaks of a *Chjon gustalinia*, i.e., a woman *gustalinia* (*mujer gustalinia*) or repeats in her rich tonal language monosyllabic words like *Xi* (*Xi santa, xi santo*), or simply uses expressions like such as *so, so, so, so* in a method of percussion with attention only to the rhythm of the canticle.

15. Mircea Eliade, *El chamanism y las técnicas arcaicas del éxtasis*. Ed. Fondo de Cultura Económica. México, 1960.

16. *Ibid.*

17. *Ibid.*

18. *Ibid.*

19. *Ibid.*

20. Sieroskewski, *Du chamanisme d'après les croyances de Yakoutes*. Cited in Mircea Eliade, *op. cit.*

Translator's notes

A. Francisco Hernández was a Spanish botanist on assignment in Mexico in the 1570's to study and describe its flora. He was one of ten primary sources cited by Wasson and Wasson (1957) for written references to the sacred mushrooms in sixteenth century Mexico. The passage Benítez quotes is part of a brief section on mushrooms found "in Volume II of Hernández' *Opera* brought out in Madrid in 1790, as Chapter 95 of Book IX of the *Historia Plantarum Novæ Hispaniæ*" (Wasson and Wasson 1957, 221).

B. Motolinía was a Franciscan friar, Toribio de Benavente, who died in 1569. Along with Francisco Hernández and Sahagún, he was another one of ten chroniclers of sixteenth century Mexico cited by Wasson and Wasson (1957) for early references to the sacred mushrooms. Among these sources, Motolinía was one of only two—the other being Sahagún—to refer to the mushrooms as *teonanácatl*, other sources using *teyhuinti* or other terms, or deriving from these earlier references. This is a point to emphasize because the term *teonanácatl* has been popularized to some extent, whereas others have not and so may be unfamiliar to the general reader. Although the term *teonanácatl* seems to have disappeared from the surviving contemporary traditions, a closely related form, *teotlaquilnanácatl*, has been reported by Gastón Guzmán in the village of Necaxa, Puebla (Wasson 1980, 44ftn.).

Motolinía's vehemence, cited by Benítez, is another distinction. Wasson states "the fury of Motolinía … against (the mushrooms) could scarce find words to express itself" (1980, 50). The friar's problem with the mushrooms seems to have been aggravated by his erroneous translation of *teonanácatl*, as "flesh of the god" (or demon). By analogy with the Christian communion services, this led him to consider the mushroom rituals were works of the devil, a mockery of Christianity as though deliberately staged to offend the Catholic conquerors. "Motolinía," notes Wasson, "… gave the word a meaning with a direct bearing upon the Elements in the Mass ... as an appalling simulacrum of the Eucharistic sacrament ..."(1980, 44). In fact, contrary to this provocative interpretation, the word *teonanácatl* really means sacred mushroom, or something very close to that ("divine or wondrous or awesome mushroom" according to Wasson 1980, 44). Motolinía's displeasure was thus largely an artifact of misunderstanding.

C. Juan Suárez of Peralta and Baltasar Dorantes of Carranza were among the sixteenth century chroniclers of New Spain born there to Spanish parents.

D. This passage apparently refers in particular to Aldous Huxley (1954), and to Antonin Artaud, author of *The Peyote Dance* (1976), which originally appeared in 1971 as *Les Tarahumaras, Tome IX, Oeuvres Complètes d'Antonin Artaud*, Editions Gallimard. The latter is seldom consulted but the former is a well known classic, widely cited as the starting point for what became the controversy surrounding psychedelic drugs.

E. References to Robert Weitlaner as "engineer" reflect a seemingly quaint Mexican custom whereby in some parts of the country, anyone wearing professional attire may be titled an engineer, a status formalized by even having its own honorific abbreviation, Ing. (from *ingeniero*). As mycologist Gaston Guzmán noted during his work in the vicinity of Rancho El Cura, near Huautla de Jiménez, he too was addressed and referred to in this fashion: "The Indians call anyone in work clothes who comes to that region from Mexico City an engineer" (1990, 94).

F. The phrase *amanita mata-moscas*, literally "Amanita kill-flies," refers to *tue-mouche,* a French folk name for the fly agaric as noted by Wasson (1968). *Tue* is the command form of the verb. In English, the owner of a dog, if he wishes it to attack, uses the common expression: "sic 'em!" The French equivalent is *tue!* meaning "kill!"(I am indebted to French scholar William Loftus for the latter information.) So this expression roughly translates into English as "kill-fly." Here Benitez is citing a passage by Heim which refers to the "Amanite *tue-mouches*." I am not aware of the expression *amanita mata-moscas* in traditional Spanish use, or other references. I assume Benitez is simply preserving the French folk name in translation for his Spanish-reading audience. As far as I know, the "kill-fly" *Amanita* is a designation with no equivalent Spanish usage.

"Kill-fly" is an intriguing apellation. "Killjoy" is the only English expression that comes to mind as possibly having a similar form, however different the meaning. The well known folk name "fly agaric" shares the fly reference, of course. And the tradition of using it to kill flies, or as though to do so, is not unknown in England. In seeing how widely scattered such expressions are among different cultures, and assuming a common origin with dispersion through time, one cannot but wonder just how ancient the traditions underlying them must be.

The term "agaric" is of Latin origin, and carries a somewhat more technical meaning than "mushroom." In modern use it is generally used in reference only to gilled fungi, or (to put it more technically) fungi with a lamellate hymenium. This generally excludes the boletes, which have pores instead of gills, as well as several other fungal forms which like agarics and boletes may have a cap and stalk, and may be casually referred to as mushrooms. Morels would be an example of the latter. Compared with "agaric" and "bolete," "mushroom" is a more recent English word derived from a term in the French fungal vocabulary (*mousseron*), and it carries no clear scientific definition. Compared with its use in America, the word is applied rather narrowly in British tradition, in reference only to species of fungi acknowledged as edible. On occasion one may see references to *Amanita muscaria* as the "fly mushroom" (for example, Kreiger 1967, Pl. 3), a translation into the American idiom, rather inconsistent with the more restrictive British notion of "mushroom." Although tenable by the American sense of the word, the idea of a poisonous mushroom would be something of a conceptual contradiction in the traditional British folk understanding. In that tradition, by definition, any fungus not acknowledged as edible would be distinguished as a toadstool, not a mushroom. The term "agaric" thus merely designates a gilled fungus, conveniently bypassing the whole question of whether it is regarded, by whomever, as a "mushroom" or "toadstool."

G. Eunice Pike did in fact write up her ethnographic observations later as she suggests she might in her letter, in a paper with co-author Florence Cowan, "Mushroom ritual versus Christianity" (Pike and Cowan, 1959). This is an important source full of fascinating information, although it can evoke a pang of irony as one realizes from reading it the extent to which the mission of the sixteenth century Spaniards, of trying to eradicate mushroom rituals among the Indians, is apparently still alive and well among at least some contemporary missionaries.

H. In the May 13, 1957 issue of *LIFE* (Wasson 1957, 102), the Mazatec name for this species was rendered as *'nti sheeto*, using a simpler notation without the numeric superscript designating pitches.

I. I have not been able to clarify *totola*, but from the context, fowl or poultry might be appropriate translations. The Wassons (1957, 258) make clear that in the ceremony described there were two turkey eggs and four chicken eggs.

J. I have been unable to clarify the term *cañuto*, and likewise failed to find any reference in Wasson and Wasson (1957) to *aguardiente* as an element of the ceremony described here on the night in question, although it figures in accounts of other mushrooms ceremonies. The other items Benítez mentions correspond closely to the Wassons' original account.

K. The *síndico* or Syndic is the office second in charge within the municipal administration, below *Presidente* or mayor. In the May 13, 1957 issue of *LIFE*, Cayetano García was referred to by the pseudonym "Filemon," much as María Sabina was called "Eva Mendez" (Wasson 1957).

L. I lack explanations for the names *Verde Machacado* (the Crushed or Macerated Green) and *Señor Estafiate*. Considering the context here, the former vaguely elicits references to a mythic-folkloric motif, apparently widely distributed, described by Carl A. P. Ruck as the Green Man, Robin of the Wood, Jack-of-the-Green, or "'Master of Animals,' the dangerous spirit whom every shaman must meet and charm: Xochipilli, amongst the Aztecs, the 'Prince of Flowers,' and every one of them that adorns his body a sacred inebriant" (Ruck et al. 2001, 4).

M. The *carrancistas* are followers of revolutionary hero Emilio Carranza.

N. The spelling of this species is given as *Conocybe siligineoides* Heim in *Les Champignons hallucinogènes du Mexique*, R. Heim and R.G. Wasson, 1958 (1959), Paris: Editions du Muséum National d'Histoire Naturelle, and other sources such as the May 13, 1957 issue of *LIFE*.

O. This apparently refers to concepts of an animal doppelganger, widespread in Mesoamerica, such as the *nagual* (or *nahual*), often conceived in malign terms; or the *tonal*, a more benign animal sign linked by one's day of birth, almost like the zodiac, but designated daily rather than monthly. Estrada (1981, 200) quotes Aguirre Beltrán to the effect that the *nagual* is an animal a person has transformed into; whereas the *tonal* has an existence separate from the person although their destinies are linked. (See in this volume, "The Mixe *Tonalamatl* and the Sacred Mushrooms" by Walter S. Miller.)

P. Flowing Water Maiden (*la Doncella Agua Rastrera*) as I have translated it seemingly refers to Woman of the Flowing Water, a figure cited by Estrada (1981, 128) in María Sabina's chanting. She is the most important female divinity of the Mazatec cosmos, a goddess of abundance and of women's work, especially that of the kitchen. In her mythic narrative, she was a young married woman who fled from her scolding, accusing mother-in-law, leaving behind the bounty of a field ripe with corn. Landscape features of the Mazatec countryside are named after events in her journey following her flight from the household of her inlaws (Estrada 1981, 227-228). Estrada (1981, 228) notes: "She is the Mazatec equivalent of Chalchiutlicue, the ancient Mexican goddess of the flowing water, 'She with the Skirts of Jade.'" I have no additional information on the references to the Lord of the Hills (*Dueño de los Cerros*), or the Horseman of Monte Clarín.

Q. This passage refers to Eliade's discussion about Eurasian shamanism and its mythic cosmos with three major layers. The world we know and live in represents the middle layer. There is an underworld below, and a celestial realm above, all connected by the *axis mundi*, the central support of the world-universe. It is understood that long ago, in the beginning, travel between these three major realms was easy for anyone and required no special skills. Unfortunately, the easy connection between them was somehow lost in a kind of fall from grace, initiating mortal human existence as we know it. Now, only shamans are still able to accomplish these mythic journeys and so they are needed to undertake them on behalf of the community and all its members.

R. I have not studied in any depth the traditions surrounding San Miguel and Señor Santiago. However, keeping in mind the cultural context of the sixteenth century Spanish Catholic influence in Mexico, the former refers to San Miguel el Arcángel, a figure widely venerated with local processions and festivals, whose name also appears as a prefix for various towns. He is known in English as St. Michael the Archangel and is mentioned in some Biblical passages, such as in Daniel 10.

The name Señor Santiago likewise has deep historic roots. It refers to Señor Santiago of Compustela, Spain, an object of religious pilgrimages going back at least to early medieval times. Santiago, which corresponds in English to St. James, also figures as a first name for some Mexican towns, such as Santiago Undameo. As detailed at a tourist information website (<www.michoacan-travel.com>), the latter town has a sixteenth century Augustinian temple devoted to Señor Santiago. Another interesting website discusses the town of San Miguel de Allende (<smaarts.com/san-miguel-9.html>), "the architectural jewel of the Americas," founded by the Spanish in 1542. This site refers to a magnificent church in a neighboring town, Neutla, which features a prominent representation of Señor Santiago high above the altar: "His image greets you from above the archway of the main entrance ... Above a side entrance ... are these interesting but undecipherable symbols which, most likely, had significance in the indigenous tradition of the area."

Catholic icons, especially as rendered in local tradition, seem to be a frequent although by no means constant feature of altars figuring in Mexican mushroom rituals. The *velada* Benítez describes here, held at the house of Herlinda Martínez Cid, must not be confused with others described and illustrated elsewhere and also led by María Sabina, particularly the services she conducted at the request of R. G. Wasson in 1955 which were made famous in *LIFE* magazine by Allan Richardson's historic photographs. Those rites, held in the home of Cayetano García and his wife Guadelupe, likewise included an altar with two Catholic icons. However, the figures depicted were the *Santo Niño de Atocha*, and the baptism of Jesus by John at the River Jordan, respectively (Wasson et al. 1974, xvi), not San Miguel and Señor Santiago. That different icons were used in these *veladas* in different homes suggests that the images and the altars upon which they stand are the property of the host, rather than the one conducting the proceedings. Excellent color photographs from the 1955 services, showing the altar in the García home with its icons, appear in Wasson (1957), Wasson et al. (1974), and Reidlinger, ed. (1990).

The baptism at the River Jordan is a well known episode from the New Testament, but the *Santo Niño de Atocha* is perhaps less familiar to many English readers, and warrants some explanation. Wasson writes:

There is a remarkable addiction among the Indians of Mesoamerica to the cult of the Holy Child of Atocha. He is, I believe, a post-Columbian embodiment of the pre-Hispanic divine *Niño*, the mushroom. The sacred mushrooms to this day are called *niños* everywhere in the highlands. In the 1950's (and perhaps even now, after 15 years of world-wide publicity and acclamation) María Sabina liked to have on her altar, as she began a *velada*, an icon of the Christ Child with his pilgrim's hat, *conchas* (bivalve shell for begging and eating), his shepherd's crook, and a bottle-gourd. The Indians in the colonial period were quick to adapt themselves to Christian practices—*fiestas*, pilgrimages, etc.— that seemed to permit an outlet for their pagan customs (1974, 134n.).

The pictorial elements identifying the Holy Child of Atocha, as described above by Wasson, are standardized to a remarkable degree. They can be seen in various depictions, for example, a painted wall mural at a housing complex of El Paso Community College (<www.epcc.edu/ftp/Homes/monicaw/borderlands/11_santo.htm>), and the website of Mr. John Todd, Jr. (<www.johntoddjr.com/36%20Atocha/at.htm>), to cite just two.

The figure of the *Santo Niño de Atocha* featured in María Sabina's 1955 *velada* at the García home looks virtually identical to similar images seen elsewhere, strikingly so, down to the most minute details of its execution, as though by photographic reproduction. For example, a book titled *Miracles of the Child Jesus* (Lord and Lord 2002) presents on its cover, in color, five variations of the Child Jesus theme, each representing a different local tradition. One of these, in the upper right, is the *Santo Niño de Atocha*. The figure as shown is removed from its background, but a more complete reproduction with the background included is also featured inside the same book (2002, 92), in black and white. The resemblance of the figure shown in Lord and Lord to the icon from the 1955 *velada* is so close that, again, it could easily be mistaken for a photograph of it. Other instances of such close, almost exact reproductions of this particular depiction can readily be seen online from search engines results using the keywords *"Santo Niño de Atocha"* (e.g., <www.johntoddjr.com/36%20Atocha/at.htm>).

However, close comparative inspection reveals small but definite differences among these images, especially in the background, demonstrating these icons are not all simply photographic reproductions. (I am indebted to John Todd, Jr. [pers. comm.] for drawing this to my attention). In fact, such icons are widely reproduced by hand and sold in Mexico, following a popular local tradition known as *retablo*. *Retablos* are "small oil paintings, usually on copper or tin, most often done by untrained artists from the provinces just outside of Mexico City," according to an internet source, "What is *Retablos* Art?" (<www.trinitystores.com/main.php4?cmd=SanterosRetablos>). In its discussion, the latter site notes that in the last quarter of the nineteenth century, "Small *retablo* factories were established ... to produce and reproduce the same images for the church, household shrines and processions. A typical '*retablero*,' rarely recognized as an artist, may have reproduced the same image hundreds, if not thousands of times in his career." The *Santo Niño de Atocha* has also been cited as a frequent subject of *retablo* depiction (<www.udayton.edu/mary/questions/yq/yql20.html>). The *retablo* tradition no doubt explains the fact of these various images of the *Santo Niño de Atocha* which appear virtually identical, as though they might be photographic reproductions, yet also reveal distinct differences, such as in the background.

The photograph presented by Lord and Lord (2002, 92) is from an image enshrined at a church in Plateros, not far from Fresnillo, Zacatecas (pers. comm.). The treatment and overall style suggest the original would be an oil painting within European tradition, presumably medieval. However, I have not been able to learn the name of an artist to whom it is attributed. The icon from the altar of María Sabina's 1955 *velada* in the García home is in all likelihood a *retablo*. The latter portrays a setting with a wooden floor, and an abstract cerulean background with distinct luminous arcs, whereas in the picture from Lord and Lord (2002), the ground is represented as a meadow with flowers, and the scene is more clearly daytime outdoors. The image posted at Mr. Todd's website is a scan of an icon purchased in the market at Plateros (pers. comm.), and features a darker, vaguely cloudy background, somewhat impressionistic, and different in its own way.

An especially diverting detail in the upper portion of the icon from the 1955 *velada* is the appearance of three beings or entities, shown as detached heads of young children hovering or floating, with vague, wing-like collars but no bodies, apparently gazing down upon the *Santo Niño*, as though tenderly. The 1955 icon also featured separate tiny scenes in its upper corners, set apart in an ornate circular border motif. This border with its miniature scenes, and the almost surreal floating infant heads, also appears in the version posted at <www.johntoddjr.com/36%20Atocha/at.htm>. But neither the border with its separate scenes nor the floating infant heads appear in the Lord and Lord photograph (2002, 92). However, the image they present appears cropped, based on the comparatively large size of the figure of the *Santo Niño* within the frame.

Not all the mushroom ceremonies documented by the Wassons and others feature iconography such as that discussed here. For example, no such portraits or images seem to have figured in the rites conducted by Aurelio Carreras, the Mazatec *curandero* who led the first ceremony the Wasson's witnessed, in Huautla in 1953. The same can be said for the Zapotec shaman Aristeo Matías, whose ritual as observed in 1954 apparently included no such altar (Wasson and Wasson 1957). However, the mushroom ritual observed in Juxtlahuaca, 1960 by Wasson and Ravicz (see Chapter Four, Ravicz, *The Mixtec in a Comparative Study* ...) included an altar with two framed icons, presumably also *retablos*. These were reported to be of the Virgin of Guadelupe, a well known icon of Mexican origin; and the Virgin of Remedies, a figure who, as explained by Margaret Houston, is venerated at the cathedral in Oaxaca City, and perhaps less widely known outside of Mexico. A photograph of the altar is featured in the original article (Ravicz 1960[1961], 81), but details of the icons shown are not distinguishable.

S. As explained by Estrada (1981) the mountain rising above Huautla, which the Spanish call the Mountain of the Adoration, is known to the Mazatec as Nindó Tokosho, and is regarded as sacred. Chicon Tokosho is a patron spirit being of Mazatec lore, reputed to inhabit the mountain. He is also known, less formally, as Chicon Nindó, meaning the Man of the Mountain (Estrada notates these names as $Chi^3con^3 To^4co^2xo^4$ and $Chi^3con^3 Ni^3nto^3$). Tokosho refers to "a type of red berries," perhaps a metonym for the mountain. "It is said that he is Lord and Master of the Mountains, that he is white, and that he has the power to enchant spirits and to exorcise evil influences or spirits that cause illness. Some identify him with Quetzalcóatl" (1981, 195).

T. *Chaneque* is one of various terms in Mesoamerica, along with others such as *enanito*, *duende*, etc., denoting native concepts, corresponding to a diverse folkloric motif widely noted across culture, of "little people," such as the leprechauns, fairy-folk, trolls and gnomes, etc. (Jacques Vallee and others have noted that various popularized conceptions of "little green men" or other diminutive extraterrestrials, as they are purported to be, may represent contemporary examples in this category.)

The term *chaneque* does not figure in other primary sources presented in this volume, but its form is strikingly similar to another term for such beings, *chamaco* (see "The Mixe *Tonalamatl* and the Sacred Mushrooms" by Walter S. Miller). A name derived from the latter term, *Los Chamaquitos*, appears in an interesting and informative article, "The use of hallucinogenic mushrooms for diagnostic purposes among some highland Chinantecs" by Rubel and Gettlefinger-Krejci (1976). The Chinantecs use the designation *Los Chamaquitos* interchangeably with *Los Niños*, "the Children" who are believed to appear and reply to questions from the one intoxicated on the mushrooms. The latter reference appears identical to that employed by María Sabina, as discussed here by Benítez. Nahuatl speakers in the Valley of Mexico also speak of the mushrooms as *los niños* of the waters (*apipiltzin* in Nahuatl, the initial *a-* referring to water), as the Wassons learned in 1954 in the village of San Pedro Nexapa, on the slopes of Popocatepetl (Wasson and Wasson 1957, 305-306).

Rubel and Gettlefinger-Krejci (1976) offer some particularly interesting observations concerning the designation by the Chinantec of some mushrooms as "boys" and others as "girls" (*machos* and *hembras*), or male and female. This tradition, with various nuances, appears in other Mexican cultures as well, especially where the mushrooms are handled or taken in pairs, with each pair representing a coupled male and female. (For the general reader, it is well to understand that, although these fungi are indeed sexually reproducing organisms—i.e., their life cycles involve fusion of fertile haploid nuclei to form a diploid nucleus—mushrooms are not of different sexes biologically in the way most animals are. In this respect they compare more closely with a plant that plays the roles of male and female by producing both pollen and seed, as is commonly [although certainly not always] the case. Indeed, furthermore, a mushroom is no more a whole, individual being or biological organism than is a flower cut from a bush; it is an organ, a part of the structure of a greater whole, the individual organism from which it came.)

Among the Chinantec, the gender of the mushroom taken, as reckoned by native tradition, is correlated with that of the little spirit who appears, i.e., whether it will be a *chamaquito* or a *chamaquita*, almost as if the little spirit is the mushroom itself, personified. Nor is one obliged to take the mushrooms *casada*, in coupled pairs, as among some of the ritual traditions documented elsewhere. In this connection, the males and females are purported to have differing tendencies that may call for deliberation, according to Rubel and Gettlefinger-Krejci's informant Don Antonio. The latter stated, for example, that "the boys are more powerful and talk more" (Rubel and Gettlefinger-Krejci 1976, 237). By eating only *machos* or only *hembras* one can have, as one wishes, a consultation with an exclusively male or female mushroom audience, although there is general agreement that a mixed group is usually best.

These considerations underscore the ritual importance of being able to tell male mushrooms from female; but how is this done? "During the collection of the mushrooms, (Don Antonio) had distinguished males (*machos*) from females (*hembras*) based on his understanding that the larger mushrooms were male" (Rubel and Gettlefinger-Krejci 1976, 240).

From the totality of the literature, it seems that ideas concerning little elf-like spirit beings associated with the mushrooms are close to the core of nanacatism in Mesoamerica, and pervasive in various forms. Furthermore, these little beings, *Los Chamaquitos* or *Los Niños*, or the *nanakatsitsen* ("little mushroom men;" see Reyes G., "An Account Concerning the Hallucinogenic Mushrooms"), are anthropomorphic, including as to their sex, being male or female. The ritual designation of each sacred mushroom as either male or female, and prescribed practices regarding their pairing for ingestion, seem to be aspects of this tradition. Considering the proposed cultural connection between nanacatism in Mesoamerica and shamanistic use of *Amanita muscaria* in the Old World, traditions of "little folk" associated with the latter species in Siberia are especially interesting (Wasson 1968, 288-301; Wasson et al. 1986, 68).

Chapter Seven
One Step Beyond: The Sacred Mushroom (ABC-TV)

One Step Beyond was a weekly half-hour television series that aired in 1959-1961, with Merwin Gerard as creator and associate producer, Collier Young as producer, and John Newland as director and host. The theme of the series was ESP, the paranormal and various other strange occurrences and phenomena. Ninety-six episodes were produced in total, and the one of interest here, *The Sacred Mushroom*, originally aired January 24, 1961, during the series' third and final season. It featured host Newland, producer Young, and special guest Andrija Puharich (Gerani and Schulman 1977, Muir 2001). The following text was transcribed from an audio cassette recording of the show, obtained in the 1970's from a broadcast by a local independent television station, plus a videotape recording of a similar broadcast, generously furnished for study courtesy of John Kenneth Muir. His book *An Analytical Guide to Television's* One Step Beyond *1959-1961* (Muir 2001) is a definitive account and critical appraisal of this classic anthology show.

One Step Beyond normally starred television actors playing roles in dramatized stories based upon purportedly true events or incidents. Occasionally the show featured a brief appearance by a real life witness or party to one of the stories, but generally not members of the crew. Newland appeared in each episode as host, but only in a brief introduction and epilogue, not as a principal player. In this regard, *The Sacred Mushroom* was a unique installment for the series, for on this one occasion, the show's production team became its stars (Muir 2001).

As Gerani and Schulman (1977, 33) note, this episode documents series producer Young and director Newland on a quest for, and later experimenting with, a mushroom that reputedly enhances extrasensory capabilities. *The Sacred Mushroom* exemplifies the subject matter of the series *One Step Beyond* in general, especially in terms of its focus on ESP and psychic experiences. But it also gives an indication of how the magic mushroom captured the interest of the general audience during a brief moment in history, a window in time immediately following the publicity of the Wasson expeditions to Mexico in the 1950's, and prior to the rise of the psychedelic sixties with the air of disrepute that came to characterize journalistic coverage of LSD and its social impact.

Director-host Newland remembered *The Sacred Mushroom* vividly in a 1999 interview by John Muir, calling it "our most popular episode ... it was pretty damn unusual to see people getting high on TV in 1961, wasn't it?" (www.johnkennethmuir.com). Newland described the journey to Mexico as "spooky ... We landed at a tiny airstrip near a mission. From there it was a donkey trip of four days to reach the village. It was a dangerous journey, but we got phenomenal footage" (Muir 2001, 189). Alcoa, the show's sponsor, objected to airing it until, strangely enough, footage was added of Newland sampling the mushrooms back in California for a study of their influence on psychic ability. "Alcoa saw it and considered my testimony 'proof enough' to air the show," recalled Newman (www.johnkennethmuir.com). Asked by Muir whether he experienced anything psychic or paranormal with the mushrooms, Newland replied: "None. Not a grain" (Muir 2001, 191). This poses a slight discrepancy from a sequence near the end of the show, in which he seems to indicate that he did indeed experience something unusual in connection with the tests for psychic ability, while Puharich, coordinating the proceedings, explains that his results are "significant for ESP."

In the first act of *The Sacred Mushroom*, show host Newland notes that the idea for the episode came from a book by Andrija Puharich published in 1959, also titled *The Sacred Mushroom*. As the title and date suggest, this was a work that surfaced in the wake of the historic 1957 *LIFE* magazine feature, "Seeking the magic mushroom," as did the show it inspired. Here we get a glimpse of the sensation stirring early on in the general audience in response to the publicized reportage of the Wasson team, by visible ripples it made in the mass media and popular entertainments.

Puharich strikes a somewhat ambiguous figure in the study of Mesoamerican mushroom rituals. There is perhaps only one major work in the classic ethnomycological literature that reports any of his observations as significant, *Nouvelles Investigations sur les champignons hallucinogènes* (Heim et al. 1967). The latter features an Appendix written by Heim, titled *Expériences de M. H. K. Puharich*. (At the end of the present chapter, I offer an English translation of this Appendix, along with some pertinent Addenda.) Wasson listed *The Sacred Mushroom*, along with a second book by Puharich (*Beyond Telepathy*, 1962, New York: Doubleday), as sources in his 1963 article "The hallucinogenic mushrooms of Mexico and psilocybin: a bibliography." Wasson organized his bibliographic references by subject headings, and he listed these two books by Puharich in a section by themselves under the heading Parapsychology. This reflects two things: (1) the "fringe" nature of Puharich's interest, which was primarily ESP and the paranormal, and (2) a certain appeal for such interest residing in the hallucinogenic fungi in Mexico, owing to their conspicuous role in shamanic divination, and seemingly, on occasion, startling anecdotal results.

In the 1960's, the perennial popular fascination with the paranormal was on the upswing, as the series *One Step Beyond* generally reflects. Obviously, such interest has never been widely credited as a genuine field of scholarly or scientific study. Although there have been efforts to establish a legitimate foundation for it (e.g., the Rhine Institute at Duke University), a questionable if not outright derisive air persists. Accordingly, the idea of a connection between the paranormal and the sacred mushroom, and Puharich's activities in this field, are matters of controversy, perhaps considered inherently exploitive by some.

Jonathon Ott, in his excellent laudatory essay on Wasson's legacy, "A Twentieth Century Darwin," makes a particularly pointed reference in this regard. He cites as "most unfortunate...the appearance of *farceurs* like Andrija Puharich...who spun absurd theories based on the Wassons' research to make a fast buck" (Ott 1990, 190). The point is clear enough, and indeed Puharich's *The Sacred Mushroom*, despite its title and vintage, is not exactly a hallowed classic of ethnomycology. Wasson (1973), in a footnote from his article, "Notes on the present status of *ololiuhqui* and other hallucinogens of Mexico," takes Puharich severely to task for slipshod methodology and credibility problems in some of his reported conclusions (see Appendix later in this chapter, "Addenda").

But from another perspective, and specifically in regard to the ESP angle, some of the circumstances the Wassons encountered in connection with mushroom rituals, and reported with all apparent candor, are rather provocative and do arouse a vague sense of unease. In 1990, Masha Wasson Britten noted:

> My parents and I were aware that the mushrooms appeared to have a certain extrasensory potential. But we did not want to publicize the fact. My father explained our reasoning ... 'I had always had a horror of those who preached a kind of pseudo-religion of telepathy, who for me were unreliable people, and if our discoveries in Mexico ... were to be drawn to their attention we were in danger of being adopted by such undesirables.' Among these discoveries, or rather experiences, was the occasion when my mother saw a city at a distance in her visions and described it to us vividly. Later, while returning on a different path through the mountains than we had taken previously, we suddenly saw Mexico City far below us, whereupon my mother said: "That's the vision." It was just as she had described it (Britten 1990, 39-40).

The divinatory forecast of the Mazatec *curandero* Aurclio Carreras in 1953 about events in the Wasson family over the impending twelve months also seemed to come true in due course to an uncanny extent (Benítez, in *The Hallucinogenic Mushrooms*, cites key passages from Wasson and Wasson [1957] recounting the latter circumstances.)

Before leaving the matter of psychic phenomena and the sacred mushrooms, it is perhaps worth noting there are a few curious accounts of mushroom poisoning cases, dating from the era preceding the scientific elucidation of *teonanácatl*, which contain some passages of possible interest in this context of paranormal-like symptoms. We can now recognize psilocybin and its related alkaloids as the probable cause in these cases, both by the symptoms described, and by the type of mushrooms implicated (*Panaeolus* species).

Research on LSD and similar compounds in the last half century or so has tended to confirm that heightened suggestibility is one common effect of these substances. As a result, anecdotal reports of psychic-like effects under the influence of such drugs are difficult to analyze when they come from subjects with prior awareness of their native use as divinatory agents or reputation for enhancing ESP. In such cases, one can never be sure of the extent to which "power of suggestion" has played a role in facilitating such an experience. But suggestion can be reasonably ruled out with subjects who logically would have had no reason to anticipate such effects, as is evident in some of the early cases of "cerebral mycetism" (as hallucinogenic mushroom intoxication came to be called in medical and mycological sources almost a century ago). In one widely cited instance, reported by A. E. Verrill (1914), the patient remarked that at one point during the intoxication, "I imagined I was able, by a sort of clairvoyance, to tell the thoughts of those around me." The way in which the subject dismisses his impression as imagination here is striking, as though his essential temperament leans more toward sobriety and skepticism than interest in ESP. Another interesting case, this one from 1949, involved symptoms consistent with a spontaneous out-of-body experience and spiritual revelation: "The man ... felt that he was passing into the next life and that he could see his own body. He stated that he realized 'we just continue to carry out our work into the next life'"(Charters 1957, 269). (For an interesting discussion of psychological factors that may offer to explain such experiences without recourse to notions of psychic powers or supernatural causes, see Neher, *The Psychology of Transcendence* [1980].)

As with so many putative contributions to a field as poorly regarded as parapsychology, works such as Puharich's certainly raise legitimate questions of sensationalism. However, there can also be no doubt that he did indeed travel to some remote regions of Mexico for the purpose of contacting shamans knowledgeable about the sacred mushrooms, and succeeded in this endeavor to some extent. Moreover, he supplied specimens for study to Heim and his associates in Paris (see Appendix). This immediately tends to place him in a light less disreputable than, say, the much more famous Carlos Castaneda, who apparently fabricated accounts of Mexican shamans and their teachings out of whole cloth and brazenly presented them as nonfiction to his credulous audience, achieving fame, fortune and adulation in the process. Puharich never achieved the kind of financial rewards the "Don Juan books" garnered for Castaneda, and he died in poverty and obscurity in January, 1995, in North Carolina. As with many people genuinely fascinated by the concept of ESP, and in contrast to various con artists who have operated as psychics, mediums, and shaman's apprentices, it seems unclear that Puharich's motives were exclusively or even primarily dishonest, no matter how confused or dubious his ideas, methods and findings.

Nonetheless, based on the quote cited above by Masha Wasson Britten, it is easy to imagine R. Gordon Wasson must not have relished the paranormal angle of Puharich's offerings to the study of the sacred mushrooms, or *One Step Be-*

yond's foray into ethnomycology for the viewing audience at home. It is striking that throughout the ethnomycological literature there is no reference to this show. For their own part, however, the creators of *One Step Beyond: The Sacred Mushroom* did apparently tip their hat to the Wassons as though to acknowledge their debt, judging by a mention briefly given in the show to a copy of *Mushrooms Russia and History* (Wasson and Wasson 1957) which host Newland finds in Puharich's home library at the beginning of the last act. Indeed, with or without the blessing of the Wasson camp, Newland, Puharich and company were in their debt as followers. The village of Juquila, where the mushroom ceremonies shown in the *One Step Beyond* episode were filmed, is a location Wasson had visited in his field studies. Furthermore, the *One Step Beyond* production crew was assisted there by a key player for the Wasson team who appears in the show, Bill Upson. He is one of several special guests given a formal introduction onscreen by the host toward the beginning. But confusingly, he is introduced as "Cuivas," despite the fact he is from then on referred to by his regular name for the remainder of the show.

Bill Upson was a linguist and missionary among the Chatino Indians, working with the Summer Institute of Linguistics. The Chatino are described by Wasson as "a linguistic enclave in Zapotec country. The language is called 'Zapotecan'" (1980, 218). Upson was credited by Wasson, along with others including Eunice Pike and Searle Hoogshagen, for important contributions to studies in Mesoamerican ethnomycology (1957, 120; 1980, 151n.). In 1956, Upson introduced Wasson, along with Allan Richardson, Roger Heim, Roger Cailleux, and Guy Stresser-Péan, to the Chatino territory, including a visit to Juquila. (Note the corresponding citation of Upson's name in the Appendix at the end of this chapter.) From Juquila, the team traveled to the Chatino village of Yaitepec where Upson lived, and he gave them lodging in his home there (Stresser-Péan 1990). Of the locale, Wasson states, "We found many species of sacred mushrooms in use" (1980, 218). However, Stresser-Péan observed: "The eight days spent in Yaitepec were interesting ... But the divinatory rituals of the Chatinos do not have the same quality of communion as those of the Mazatecs, nor are they so noteworthy ... The ceremony we witnessed during an afternoon left Mr. Wasson disappointed" (1990, 233).

Further background on *One Step Beyond: The Sacred Mushroom* appears in *Memories of a Maverick* by H. G. M. Hermans (1998), a book credibly purporting to be written by a former wife of Andrija Puharich and mother of two of his children. The text can also be found posted on the internet (go to <www.uri-geller.com/books/maverick/maver.htm>). According to Hermans, Puharich learned in 1955 about the Wassons' research in Mexico, and first went to Juquila in June, 1960 with a group of nine people. In a letter dated July 16, 1960, Puharich wrote of making contact with a *brujo* named Macedonio who agreed to perform a mushroom ceremony for him. This would no doubt be the same *brujo* referred to as Macedonio in the show, one of two who appeared. One quote from the letter seems to suggest that the species involved must almost certainly have been *Psilocybe cubensis*, based on its habitat: "Macedonio then gave us a special privilege of seeing where the sacred mushroom grows. Guess where? It grows in burro corrals, right out of the dung!" (1998, 80). Other species were apparently also known to Puharich's informants, judging by another reference in the same letter, to "an old Indian named Valso who has some special large mushrooms that he gets from the mountains five hours away" (1998, 81-82). Another quote from Puharich's letter indicates the development of the *One Step Beyond* project was already underway: "My present plan is to stay on here and see if I can't work up to getting in on a good ceremony so that Johnny Newland can come down here and make a movie" (1998, 82).

Puharich eventually returned from Mexico, but not for long. As Hermans notes:

> The discovery of the 'sacred' mushroom and its effects had inspired Dr. James Dill, head of the Department of Pharmacology of the University of Washington, to sponsor and partially finance a 14-man scientific expedition to Juquila ... Included in the expedition were television series producer Collier Young and host-narrator-director John Newland. The 'One Step Beyond Company' and the sponsoring Alcoa Corporation financed all the filming ... The documentary—showing the expedition party locating the mushroom through a Mexican *brujo*, and the ESP tests before and after eating the mushroom, which were conducted at our home—was subsequently shown on ABC-TV's *One Step Beyond* (1998, 82-83).

Hermans (1998) also gives further background on another figure who briefly appears in *The Sacred Mushroom*, Deane Dugan, who is identified in the show as a missionary nurse. Puharich's letter of July 16, 1960 refers to Dugan, revealing she assisted him on his first trip to Juquila as translator and guide, introducing him to people and helping him find a *brujo* to conduct a mushroom ceremony for them. When the ceremony was held, "Deane decided to take the mushroom so that I could do the filming. The ceremony went fine, but had no effect on Deane because she vomited up the horrible tasting mushrooms as soon as she had eaten them" (1998, 81). David Gray, another principal who appears in *The Sacred Mushroom*, as "one of the last kahunas of Hawaii," is also discussed in *Memories of a Maverick* where he emerges as an individual of remarkable qualities. Gray (his name appears in *Memories of a Maverick* as Bray) later hosted Puharich on a visit to Hawaii to hunt mushrooms.

One can only imagine a show such as *The Sacred Mushroom* could never be made today, for reasons too numerous to count and, in some cases, depending upon one's perspective, perhaps too lamentable to belabor. It must have raised eyebrows in its day, and certainly held the audience's attention with its sensational scenes of Mexican mushroom ceremonies and divination, not to mention the show's host eating the mushrooms in a comfortable private setting back in California. Here, ABC-TV must be duly credited for an exceptional moment in broadcast history. The transcript follows.

*   *   *

Studio: closeup of dried fungal specimens

HOST (John Newland):

This member of the mushroom family, this fungus, is known for the moment only as X. It was discovered barely weeks ago growing in a remote rainforest. Science has not yet given it a name, for science knows scarcely anything about it. But it is felt that X might have one remarkable quality: that it stimulates extrasensory perception, enabling the mind to become telepathic and clairvoyant. Now that's a rather large claim. But is it true or false? The answer to that question took us on a unique and distant journey.

Location: Mexico (Mitla, Oaxaca)

HOST:

We're in Mexico, thousands of miles from our Hollywood soundstages, in a place called Mitla, on the very edge of civilization. There are no actors, there is no script. For this psychic report can only be recorded at the moment that it happens. Here in the plaza of the kings of Zapotec, here in the antiquity of Mexico, we begin a search whose ending at this very moment of filming we do not know. We are searching for something far older than these ruins which, if found, could hurtle ordinary man infinitely beyond his five senses. And now, another step beyond. The first step is a faltering one. We wait day after day for perfect flying weather. For the only way to reach our goal, the inaccessible village of Juquila, is by light plane over trackless mountains.

YOUNG:

Have they come back yet?

WOMAN:

No.

YOUNG:

Well, keep your fingers crossed.

HOST

This lovely *hacienda* at the foot of the (inaudible) Mountains has been here a little longer than we have, but not much. From here, weather permitting, our chartered plane will fly us to a remote mountain village to look for, of all things, a mushroom—but a very strange mushroom indeed, with powers beyond belief. And here are the mushroom hunters, as improbable a group as you will find anywhere. For instance this is Dr. Barbara Brown, brilliant neuropharmacologist of the University of California at Los Angeles. With her is David Gray, one of the last kahunas of Hawaii, a spiritual leader whose line goes back nine hundred years, and whose quest for knowledge beyond the world of materialism has led him everywhere in the world, and now here. And here with a friend (Ed.: a kinkajou or ringtail) is Jeffrey Smith, a distinguished Professor of Philosophy and Humanities at Stanford University. And now, and mighty important, is our translator, "Cuivas" (?), a missionary who has been among the people of our remote village for the last nine years, and the only man in the world who speaks their language.

PUHARICH:

The pilot says he'll be ready soon John.

HOST:

And last, our forever optimistic Dr. Andrija Puharich, who wrote the astounding book *The Sacred Mushroom*. Dr. Puharich's book brought us here in the first place, and in it he explains how the mushroom seems to have an incredible effect upon the power of extrasensory perception.

PUHARICH:

Incredible is the word. We're here for some very specific purposes. We want to know why these people here in Mexico have kept this rite so secret for so many centuries. But I would say that our prime mission is to explore and examine the biochemicals present in these new mushrooms and find out if they can be of benefit in this problem of mental disease.

HOST:

What Dr. Puharich hasn't told you—it's not all as simple as it sounds. The last time he was here, looking for the mushroom, a man was shot.

PUHARICH:

And shot at.

HOST:

And another man was driven out of his mind.

PUHARICH:

For a while.

MAN:

El piloto dice que ya.

HOST:

In English, *dice que ya* means the pilot said now.

The shadows of our two small planes fall on rainforests which hide villages where the twentieth century does not exist. And that's the landing strip, chopped by hand out of the mountain side so that missionaries can bring the word of God to this lost world. As we approach, the pilot warns us to grab hold of something. A sudden gust of wind could smash us against the mountain, or plunge us into the abyss below.

At last we are in the village of Juquila, slogging through mud, each of us with an odd sense that all the clocks have stopped ticking. We've brought television cameras to people who never even heard of radio, who never saw electric light, never heard of aspirin, let alone penicillin. But perhaps the sacred mushrooms have given them a psychic insight far beyond our deepest explorations. (station break)

Our party breaks up to go about its various investigations. Dr. Brown and a missionary nurse named Deane Dugan go looking for the mushroom guarded by armed soldiers because in past months shots have been fired, machetes have been flashed, mushroom hunters who came before us have been murdered. Dr. Puharich and missionary Upson begin searching the town for a *brujo*, a priest of the mushroom cult. But everywhere they go, an iron curtain of silence comes clanging down. Amid much shaking of the head and wagging of the finger, everyone emphatically denies the existence of a mushroom cult, and it is made very clear that if this is why we have come to Juquila, we are not welcome.

Our search seems hopeless until Dr. Puharich gets an idea that might help us push open an ancient door. Bill Upson is sent through the village announcing that a free medical clinic has opened to all. Within minutes, the first patient appears, a little Chatino Indian with a lot of aches and a lot of pains. And then they come in droves, all of them humble, all of them hopeful, all of them grateful.

Finally the mother of a child whom Dr. Puharich has treated timidly suggests that she might help. Kindness and a bottle of vitamins have worked a magic of their own. Sometime tonight we will have a visitor. Thanks to the help of the sick child's mother, secret arrangements are made, and we are guided to this hidden place to wait.

It is almost midnight before the *brujo* appears. He has practiced the secret and all but forbidden rites of the sacred mushroom for decades, rites passed on from father to son for perhaps four thousand years. But of course he has never performed before such an audience, or a camera. Our scientists have prepared test questions to evaluate the *brujo's* extrasensory perception. But he insists upon one condition: to achieve spiritual union, they also must eat the mushroom. However, in the interest of detached scientific observation he agreed that Dr. Brown does not have to take the mushroom.

The rites begin. A strangely aromatic root is burned, inhaled. The *brujo* murmurs incantations in Chatino, a language unknown anywhere else on this earth. Then our group waits for the mushroom to take effect. Bill Upson questions the *brujo*. Because of the circumstances, the sound is of course far from perfect.

UPSON:

I've just asked him when we'd see something, when something would happen. And he held up his hand to indicate he didn't want to talk just now.

HOST:

And then silence again. And then, oddly enough, the first sound we hear as the chemical in the mushroom takes effect is laughter. The *brujo* has just told of an amusing moment from Bill's childhood in Indiana—Indiana, a place the *brujo* does not even know exists. Now the group tests the extrasensory powers of the *brujo*. Sometimes he is accurate, sometimes he is not.

UPSON (translating):

—the rabbit will brighten the eyes.

PUHARICH:

(inaudible, mixed with woman's voice) ...this means clairvoyance?

HOST:

The following is one moment when he is startlingly correct. Suddenly he turns to Dr. Brown, whom of course he has never seen or heard of before.

UPSON (translating):

(inaudible) ...you're still sick, it's in your chest, your heart, your heart is sick.

HOST:

Without a moment's hesitation, he has accurately diagnosed a personal illness that only Dr. Brown knows about. Our next experience with a *brujo* and the sacred mushroom comes quickly and unexpectedly, and this time it has nothing to do with scientific investigation. Since it is now common knowledge that we have already witnessed the rites, we are allowed to come along to see a second priest at work, a man named Macedonio, who agreed to take the mushroom and listen to a villager's problem.

UPSON:

He said his burro was stolen and he wants to know where to find it.

HOST:

In this village of Juquila, the theft of a burro, a man's most important possession, his donkey, is second only to murder.

UPSON:

He says three men are involved.

HOST:

Macedonio hardly paused a moment before he named the thieves.

UPSON:

Gustavo said he was going to take the three of them to jail tomorrow, but Don Macedonio warned him not to go near the house of Juan and Pedro because they would kill him, but to look for him at the house of José.

HOST:

We go with the villagers through a valley, across fields past other villages to the exact spot where Macedonio said the burro would be. And here indeed it is, its identifying brand clearly marked on the burro's flank. Now perhaps we can understand all of this, why for four thousand years, and from one civilization to the next, the sacred mushroom has endured. We have been given a practical demonstration of how it works in the daily lives of this primitive people.

Location: Northern California

HOST:

Three weeks have passed, and we're a long way from our remote mountain village of Juquila in Mexico. We're also a long way from the secret rites of the sacred mushroom cult. But our search for the truth about the mushrooms, that is the truth as we understand it, is far from over. These men are our technical crew who have come with me from Los Angeles to help us continue our search.

We've come to the home and laboratory of Dr. Andrija Puharich in northern California. For me personally this next step beyond may be a rather large one (station break).

This is part of the laboratory and study of Dr. Puharich. Here among the solid and sound books of orthodox medicine there are also some other books. They are written in French, in Spanish, in English, here is one that says *Mushrooms, Russia and History*. They're written in Syrian, Japanese, all the languages of the world. And they're all concerned with one thing, the riddle of the sacred mushroom. Now when we were in Mexico, the *brujo* who discovered the stolen burro might have been coincidence, and the other *brujo* who told Dr. Brown about her past illness might have made a wild guess. But today we should be able to prove the case for or against a mushroom, the sacred mushroom, with quite a lot more accuracy.

PUHARICH:

All set now John.

HOST:

Are these the mushrooms you brought back from Mexico?

PUHARICH:

Yes.

HOST:

Andrija, how many people have you tested since you've been back?

PUHARICH:

You'll be number twenty-six.

IIOST.

How about the results?

PUHARICH:

Encouraging. The average of the group as a whole, and the series is quite small, you understand, shows evidence for extrasensory perception after taking the mushrooms.

HOST:

Well I'm ready, that is, ready as I'm ever going to be.

PUHARICH:

Well, not quite yet. We have to do some control tests first to see if you have any ESP before you take the mushroom.

HOST:

Oh, I see.

PUHARICH:

And these will be done with cards, pictures, and books. Now you just sit still there.
HOST:

Who knows, I might turn out to be a great big psychic without anybody's help from anything, mushrooms—where are you?

PUHARICH:

I'll be with you—excuse me. Now John, you take your left hand and bring it in here on this first block. There's a row of ten blocks by you. You bring your right hand over and you scan this far row and you remain sensitive to any impressions. And when you think you're over the card that matches the one your left hand is on you pick it up and move it opposite that card.

HOST:

I see. It's like tall buildings in Chicago or New York or somewhere. Well, how did I do? (sigh of disappointment) Tall buildings in Chicago! That's not—but how did I do with the rest of the—

PUHARICH:

Well, I've got your scores added up. On this test, you got five hits out of fifty. On the book test here you got one out of sixteen, and you get nothing on the pictures. This is a chance score, and shows—

HOST:

It shows that despite three years with our show I have what they would call practically no extrasensory perception at all.

PUHARICH:

That's what the tests show.

HOST:

Well, what do we do now?

PUHARICH:

Now, to eat the mushroom.

HOST:

All right, I'm ready, I'm ready. Which ones are we going to eat? You have so many here Andrija. Which ones—?

PUHARICH:

These are for you. Chew them well when you ingest. Not bad, are they?

HOST:

Tastes just like mushrooms.

PUHARICH:

And one more for you.

HOST:

All right, what now?

PUHARICH:

Now we'll lie you down on the bed.

HOST:

What for?

PUHARICH:

We'll do some physical testing before the effects of the mushroom take hold.

HOST:

Like what? And like what effects?

PUHARICH:

You'll find out. (time lapse) Well it's been about twenty minutes since you had the mushrooms John. How do you feel?

HOST:

Fine. I feel—strong, a sense of well being.

PUHARICH:

Yes, some people feel that way. Let's try something. I want you to stand right up here in front of this light, close your eyes, and tell me what you see. Lower your head a little bit. A little more.

HOST:

I see, now I see, I see so many things I can't tell you. Can't you slow it, slow it, now I see it, so many colors that are—

PUHARICH:

Are they pretty?

HOST:

—and geometric signs, now it's like, they're magnificent, they're magnificent, I've never been so aware of color, of color, I feel like, I would like to dive into the middle of it.

PUHARICH:

Well, we'll try the card test again. All right, same way, start with your left hand and try to find the one in this row.

HOST:

(inaudible) like the dunes of a desert or like half hoops (inaudible).

PUHARICH:

That's (inaudible). Want to try another one?

HOST:

It has a great deal of, it has something of power, like noise. I don't know how to translate that into a picture. It seems to have noise and speed. There's a face looking this way, now a face looking that way.

PUHARICH:

That's pretty good, would you like to take a look and see what it's like? (pause) Amazing, isn't it?

HOST:

Before you tell me anything about that score, let me tell you something, that I did feel something sort of profound—no, profound, when touching the picture, when touching the cards I felt something, I don't know how I did but the sensation—

PUHARICH:

Well, John, I can tell you, it wasn't a delusion on your part because your card scores were twelve hits out of fifteen, which are significant for ESP. You got three out of sixteen page numbers in the book, and of course, most intriguing of all was your work with the pictures, you got six adequate descriptions out of eight pictures.

HOST:

You know Andrija I think I understand a little why for centuries they have called this mushroom sacred.

Studio

HOST:

Now our mushroom hunters are reassembled here in our studio. Dr. Brown was our search worth it?

BROWN:

Yes, I think so. The chemicals in the mushrooms, I think, will offer us another tool by which we can explore the mechanisms and the capabilities of the mind of man, and that's both sensory and extrasensory.

HOST:

What about you Professor Smith?

SMITH:

Well John, the more I experience in life, in its ordinary and extraordinary aspects, the more I feel like a wide-eyed child reaching for knowledge and life. As for the mushroom, it took me out into worlds I didn't believe existed, far beyond ordinary perception. And yet even today, they're as real and as important to my everyday life as anything I've ever experienced.

HOST:

Dr. Puharich?

PUHARICH:

During our stay in Juquila, I witnessed quite a number of cases of extrasensory perception by the *brujos*, some of which we could not film as you know John. This, in addition to the ones we did film, and the work which we've done since our return, convinces me that the mushroom contains a chemical which is most promising for the further investigation of extrasensory perception.

HOST:

When this program began, the question was, were the claims for the mushroom true or false? Well for those of us who made the journey, the answer is: true.

*   *   *

APPENDIX

Experiences of Mr. H. K. Puharich, translated from *Nouvelles Investigations sur les champignons hallucinogènes*, Heim et al. 1967, p. 219. (Ed.—There is no doubt this refers to Andrija Puharich, despite the difference in his first name as given in this source.)

> Some information worthy of mention was communicated to us by Mr. Henry K. Puharich, M.D., of Carmel, California, after his return in 1960 from a stay of two months in the Chatino country, in northwest Tehuantepec, close to the Pacific coast.
> One of these mushrooms, not identified as yet in the absence of samples and spores, appears different from all the agarics with pigmented spores that have been described as species with psilocybin: *Psilocybe, Stropharia, Conocybe, Copelandia,* and *Panaeolus.* The Chatino name for this species would be

kwi ya hojo kwitsi, meaning "sacred rabbit." The pileus, which is completely scaly as is the stipe, measures 4 to 5 cm in diameter. It appears in summer "in the marshes and very wet places." The average dose corresponds to a pair. It acts in a powerful manner presenting a very potent effect and "makes it possible to see far beyond the horizon." Mr. Puharich consumed two specimens. The first hallucinations appeared at the end of one hour in a form like the cells of a hive, then a paved road, narrow, bordered by high stone walls, which wound themselves around the experimenter. Architectural representations, of place— cathedrals, mosques—apart from any living being, intruded into the visual scene. Then "a dizzy spell broken by happy smiles appeared." Then, the hallucinations appeared in forms "watermarked," like baskets or braided nets or mesh wires. An hour and a half after the beginning of these various demonstrations the intoxication was finished leaving the "patient" with a sense of pleasant relaxation. The diameter of the pupil had gone from 3 to 7 mm. Neither the pulse, nor the blood pressure were affected except, as to the former, during the last stage of these demonstrations when it dropped from 72-76 to 52 beats per minute. After sleeping, Mr. Puharich felt, all day long, an exceptional feeling of well-being. These various symptoms, somatic and psychic, no doubt present a similarity to those which several experimenters felt with "teonanacatl" including myself by the end of studies undertaken among the Indians in Huautla de Jiménez, in 1956.

Another experiment was attempted with a species which appears to me most likely identifiable as *Psilocybe zapotecorum* (*kwi ya ho o*). The hallucinations, not of the geometric type, appeared in "pastel colors, dull, following serene landscapes."

Lastly, another trial took place with *P. zapotecorum* var. *elongata*, carrying the names *el hombre* and *nu eu naha kwi ya*, and coming from Panixtlahuaca; larger specimens are probably still referable to *zapotecorum*; they are called *mujer* and *kwi ya*.

According to Mr. Puharich, the *mujer* would be consumed by the men, and *el hombre* by the women, the amount still being a pair. For this experimenter, the action of *el hombre* was not so much hallucinogenic as toxic in nature, producing a strong intoxication, accompanied by a dysphoric reaction, characterized by fear. No disturbance appeared in the pulse, the blood pressure or even in the pupil. A deterioration of the colors occurred however (Ed.—I assume this refers to facial pallor). A tendency toward a paranoiac state was also recorded. Mr. Puharich again insists on a sense of unusual strength which he felt the following day during the ascent of a neighboring mountain, El Cerro Grande, which rises to more than 3000 meters. This observation would remain as the only beneficial result of this new trial, which otherwise produced "unpleasant" psychic effects.

These are the new data from the Chatino country the consumption of the hallucinogenic mushrooms led our correspondent to after the proper investigations we pursued in 1958 in the same region with R. G. Wasson, G. Stresser-Péan, and B. W. Upson. We have mentioned in addition (p. 160) the determination of the el hombre as a variety of *Psilocybe zapotecorum*, the var. *elongata*. But our opinion, which would deserve to be confirmed, remains that all of these mushrooms belong to the species *zapotecorum*, except probably the "sacred rabbit" whose identity remains mysterious. In any case, this information supports the presence and use of *teonanácatl* in the Chatino country, as well as the watery habitat, large size, and the particularly active effects of *P. zapotecorum*, whose place among the old rites still surviving, related to the ingestion of the sacred mushrooms, remains dominant in this province as at the turn of the century—today on the verge of disappearing—in Totonac country.

R. H.

Addenda

1.

The reference by Heim in the passage above, to page 160 and the identification of the *el hombre* mushroom Puharich reported, prompts me to add the following additional text, translated from the same source, (Heim et al. 1967), pp. 160-161. This portion is co-authored by Heim and Cailleux and, as these authors are both "double-aught" mycologists, there is a bit of technical terminology. Some of it is obvious, such as references to the carpophore (mushroom), or the pileus (cap) and whether or not it has a mammillate (nipple-like) umbo (central protrusion), or the stipe (stalk) and whether it has an annulus (ring). Some of it is more subtle, such as reference to the *type* of a species. For the record, the type is a voucher collection, scientifically documented and deposited in an accredited herbarium following proper procedures, upon which the published description and concept of a species is based.

One thing the following source illustrates is that Puharich apparently did obtain collections and send them for competent study to the top authorities in the field, Wasson's French mycological associates Heim and Cailleux, who duly identified them, as set forth in the text which follows:

Psilocybe zapotecorum Heim: New Wild Forms

The species *zapotecorum* was later reported in the Totonac country and Chatino country, respectively by Mr. Guy Stresser-Péan in the area of Misantla (1959) and by Mr. H. Puharich during his trip of July through September, 1960, to the Juquila area where one of us (R. H.) had already gone with R. G. Wasson in 1958. Measurements of the associated spores yielded the following results:

Misantla: (5.2) 5.5-7 x 3.5-4.3 x 3-4 μm
Juquila: (5.2) 5.5-7 x (3.2) 3.5-4 x 3-3.5 (4) μm

In other words, almost identical to measurements previously obtained (Heim and Wasson 1958, p. 149) but nevertheless a bit smaller. The elongation of the samples from Misantla and Juquila prompted one of us (R. H.) to describe it as a distinct form, *elongata* Heim (Plate VII), the diagnosis of which is as follows:

Form differing from type collection by the comparatively long stipe (3.5 to 5 times the width of the pileus), and by having a well differentiated, mammillate umbo at the center of the pileus. Spores 5-5.5-7 x

3.2-4.3 x 3-4 µm. In watery or very humid places. Territories of the
Chatino, and Totonac, Mexico.

Let us add from the letter Mr. Henry K. Puharich wrote to us on October
12, 1960, some useful details he brings to bear on this form. The pileus of the
mushroom measures 12 cm in diameter, and the length of the stipe reaches 18
cm. The surface of the pileus is slightly viscid, glabrous, and the color of "dark
green algae;" that of the stipe is clear green. Both pileus and stipe have a "rub-
bery" consistency, firm in the fresh state; "to the touch, they make me think of
marine algae." This form "sprouts in the marshes," which confirms its identification and at the same time our affirmations as to the aquatic nature of this
mushroom. In Chatino, according to Mr. Puharich, it retains the rather general
name of *kwi ya ho*, close to a designation previously assigned to *P. zapotecorum* and *caerulescens*. This is the form to which specimens from Panixtlahuaca (collected by *brujo* Mancillo) are assigned and they would resemble *P.
caerulescens* somewhat, if the spores and habitat did not make it possible to refer them back to *P. zapotecorum* var. *elongata*. Its Spanish name is *el hombre*,
and Chatino, *ne kwi-ya*. During their expedition in 1956, R. Heim and R. G.
Wasson had gathered the Chatino name *cui³ya²jo³o³ ' tnu³* meaning "big sacred
mushroom," a name very close to the one used to designate the variety *nigripes*
of *P. caerulescens*: *cui³ya²jo³o³ su⁴* meaning "sacred mushroom of power." The
usage of this form is reserved for women. This is the one with which Dr. Puharich reported the very acute intoxicating reaction to which we also refer (p.
219, "Experiences of Mr. H. K. Puharich").

2.

The following quote by Wasson is noteworthy in connection with his inclusion
of two books by Puharich (*The Sacred Mushroom*, 1959 and *Beyond Telepathy*,
1962) in his bibliography (1963) of published sources on the hallucinogenic
mushrooms. This passage occurs as a footnote in Wasson's article, "Notes on
the present status of *ololiuhqui* and the other hallucinogens of Mexico," which
first appeared in 1963 (Botanical Museum Leaflets Harvard Museum 20: 161-
193). It was republished in 1966 under a shortened title in *Summa Antropológica
en Homenaje a Roberto J. Weitlaner* (the same volume in which Walter S.
Miller's article on the Mixe *tonalamatl* and the sacred mushrooms was published). It is excerpted below from a later reprinting (Wasson 1973).

As we have seen, Puharich obtained collections from the Juquila region that
proved to be of sufficient interest for Heim to report on in his work on the sacred
mushrooms. Here, Wasson gives some specific reasons Puharich's other ethnomycological contributions are not more highly regarded, at least by some, even
apart from any questionable emphasis upon a psychic or paranormal research
angle:

> In *Beyond Telepathy* (Doubleday, N.Y., 1962) Andrija Puharich on p. 20 had
> written, 'The author was also informed by certain *brujos* among the Chatino
> Indians (living in Southern Oaxaca) that they used the *Amanita muscaria* for
> hallucinogenic purposes. The proper dose is one-half of a mushroom.' If true,

this would be sensational. It is not true. *A. muscaria* is the hallucinogenic mushroom of the Siberian tribesmen in their rites. It is not used in Mexico. When we first began visiting the Indian country of southern Mexico, we were expecting to find that the hallucinogenic mushroom there was *A. muscaria.* For ten years we combed the various regions and we have invariably found that it played no role in the life of the Indians, though of course it is of common occurrence in the woods. We had visited the Chatino country, where we were accompanied by Bill Upson of the Instituto Lingüistico de Verano, who speaks Chatino. Later he likewise helped Puharich, but he informs us that no *brujo* in his presence testified to the use of a mushroom answering to the description of *A. muscaria.* After the Puharich statement had appeared, I gave Bill a photograph in color of *A. muscaria,* and he returned to Juquila and Yaitepec. An informant named Benigno recognized the mushroom at once and identified the stage of development that it had reached, as would be expected of a countryman intimately familiar with his environment. He said the people in his area do not take that kind of mushroom. Chico Ortega is a Zapotec Indian of mature years, keen intelligence, high sense of responsibility, and vast experience throughout the villages of the State of Oaxaca. In the summer of 1962 I sent him, with the color photo, to sound out Chatino villagers as to the use they made of it. Discreetly, he went from village to village. The results were uniformly and unanimously negative. Puharich in *The Magic Mushroom (sic)* as well as in his most recent book is unduly impressed with the occurrence of *A. muscaria.* Wherever the species of trees occur with which it lives in mycorrhizal relationship, it is common. It is one of the commonest of fungi in North America and Eurasia. Puharich quotes at length as an authority Victor Reko, a notorious *farceur,* not to be confused with his cousin Blas Pablo Reko. Puharich did not identify the spot where he met his *brujos,* though it seems probable that he did not get beyond the *mestizo* town of Juquila. He does not explain how he put his question to them, how he explained over a double linguistic barrier what *A. muscaria* looked like. He does not explain what precautions he took to avoid a leading question that would almost certainly produce the desired answer (Wasson 1973, 188-189 [note 27]).

Glossary

aguardiente: distilled sugar cane liquor, analogous to "moonshine."

angelito(s): "little angel(s)," *Psilocybe mexicana* Heim, also known as *pajarito(s)*; or *pajaro(s)*.

amate: Mexican bark paper, a traditional pre-Columbian product; from Nahuatl *amatl.*

atole: a traditional *masa* beverage, almost like porridge or thin gruel, served warm.

bagasse: rotting plant debris, especially of sugar cane.

bruja/brujo: witch doctor, shaman in the broad sense, of lower regard than *curandero* or *sabio* (*embrujar*, to bewitch).

brujeria: witchcraft, malicious use of supernatural means, especially by a *brujo/bruja.*

chamaca/chamaco: dwarf or elf, little beings of folklore.

chaneque: dwarf or elf, little beings of folklore.

compadrazgo: godparentage.

comadre/compadre: godparents.

comal: round griddle for cooking tortillas.

copal: indigenous incense in Mesoamerica, a fragrant, resinous exudate of certain tree legumes, especially the genera *Copalifera* (literally "copal bearing," obviously named after its native use) and *Hymenaea.*

curandera/curandero: native healer combining empirical and mythic-ritual methods.

curanderismo: practice of the *curandero/curandera.*

derrumbe: "landslide," a name for *Psilocybe caerulescens* Murrill.

desbarrancadera: "landslide," a name for *Psilocybe caerulescens* Murrill.

duende: spirit being.

dueño: supernatural or elemental being, patron or lord of a place or natural domain.

elote: sweet corn.

enanita/enanito: dwarf or elf, little beings of folklore.

escamole: reputedly the tastiest of the insects traditionally eaten in Mexican cuisine, immature stages of certain ants.

hechicera/hechicero: witch.

huipil: traditional blouse worn by indigenous women in Mesoamerica.

jilote: ear of corn.

limpia: cleansing ceremony or ritual, against bad spirits or witchcraft.

mano: hand stone, used with *metate* for grinding, especially food preparation.

masa: ground corn.

mayordomia: religious society that sponsors celebrations for the church in Mesoamerica;

counterpart of pre-Columbian organizations, official appointments to which are
called *cargas* or cargoes, i.e., burdens, the highest of which is *mayordomo.*

mezcal or *mescal:* distilled liquor made from maguey (blue agave); tequila for example.

metate: grinding stone, used with *mano* or hand stone.

milpa: corn field.

nagual (nahual): sorcerer, especially in shape-shifted form as an animal alter ego.

nanacate: Mexicanism for "mushroom" derived from Nahuatl *nanacatl.*

nixtamal: (dried maize, treated with lime and partially cooked).

Nahuatl: language of the Aztecs and others in their linguistic branch, collectively called
the Nahua.

ololiuhqui: seeds of morning glory species containing lysergic acid amide, used as
hallucinogens in Mexico, sometimes in place of sacred mushrooms during the dry
season.

olote: corn cob (minus kernels).

pajarito(s): Psilocybe mexicana Heim, also known as *"angelito(s)"* (form usually plural).

petate: floor mat.

pisiete: indigenous tobacco.

Popol Vuh: a sacred Mayan book, telling the story of two brothers on a mythic quest.

pulque: fermented drink made from maguey (*Agave* sp.).

sabía/sabío: wise one, term of great respect for a native healer.

sabiduría: practice of the *sabía/sabío.*

sarape: traditional shawl.

tamale: a popular dish, vegetables and meat encased in a shell of *masa* dough, wrapped
in a corn husk and steamed or boiled.

tona: mythic animal born at the same time as a human, with fate linked accordingly.

velada: term used by Wasson and others for the mushroom ritual, as a nightlong vigil.

Bibliography

Allegro, John M. 1970. *The Sacred Mushroom and the Cross*. Garden City, New York: Doubleday.

Anderson, Edward F. 1980. *Peyote: The Divine Cactus*. Tucson: University of Arizona.

Anonymous. 1973. Psilocybin demand creates new drug deception. *PharmChem Newsletter* 2: 1-2.

Artaud, Antonin. 1976. *The Peyote Dance*. New York: Farrar, Straus and Giroux.

Benítez, Fernando. 1970. *Los hongos alucinantes*. In *Los Indios de México, Libro I, Tierra de brujos, III*, pp. 205-282. México DF: Biblioteca ERA.

Borhegyi, S. F. de. 1961. Miniature mushroom stones from Guatemala. *American Antiquity* 26: 498-504.

Braden, William. 1967. *The Private Sea: LSD and the Search for God*. New York: Bantam.

Britten, Masha Wasson. 1990. My life with Gordon Wasson. In *The Sacred Mushroom Seeker*, ed. Thomas J. Reidlinger, pp. 31-42. Portland, Oregon: Dioscorides Press.

Charters, A. D. 1957. Mushroom poisoning in Kenya. *Transactions of the Royal Society of Tropical Medicine and Hygiene* 51: 265-270.

Conconi, Julieta R.F. 1982. *Los Insectos Como Fuente de Proteinas en el Futuro*. Edit. Limusa.

Dow, James. 1986. *The Shaman's Touch: Otomi Indian Symbolic Healing*. Salt Lake City: University of Utah Press.

Earle, F.S. 1906. *Algunos hongos Cubanos*. Primer Informe anual, Estación Central Agronómica de Cuba, Habana.

Escalante H., Roberto, and Antonio López G. 1972. Hongos sagrados de los matlatzincas. *Proceedings (40th International Congress of Americanists)* 2: 243-250.

Estrada, Álvaro. 1981. *María Sabina: Her Life and Chants*. Transl. Henry Munn, Santa Barbara: Ross-Erikson.

Evans-Pritchard, E. E. 1937. *Witchcraft, Oracles and Magic among the Azande*. Oxford: Clarendon.

Furst, Peter T. 1972. Peyote among the Huichol Indians of Mexico. In *Flesh of the Gods: The Ritual Use of Hallucinogens*, ed. Peter T. Furst, pp. 136-184. New York: Praeger.

Furst, Peter T. 1974. Hallucinogens in pre-Columbian art. In *Art and Environment in Native America*, ed. Mary Elizabeth King and Idris R. Traylor, Jr., pp. 55-102. Spe-

165

cial Publications of the Museum Texas Tech University, No. 7. Lubbock: Texas Tech Press.

Furst, Peter T. 1976. *Hallucinogens and Culture*. San Francisco: Chandler and Sharp.

Gerani, Gary, and Paul H. Schulman. 1977. *Fantastic Television*. New York: Harmony Books.

Guzmán, Gaston. 1978. Further investigations of the Mexican hallucinogenic mushrooms with descriptions of new taxa and critical observations on additional taxa. *Nova Hedwigia* 29: 625-644.

———. 1983. *The Genus* Psilocybe: *A Systematic Revision of the Known Species Including the History, Distribution and Chemistry of the Hallucinogenic Species*. Vaduz: Beih. Nova Hedwigia 74, J. Cramer.

———. 1990. Wasson and the development of mycology in Mexico. In *The Sacred Mushroom Seeker*, ed. Thomas J. Reidlinger, pp. 83-110. Portland, Oregon: Dioscorides.

Heim, Roger, Roger Cailleux, R. G. Wasson and P. Thévenard. 1967. *Nouvelles Investigations sur les champignons hallucinogènes*. Paris: Èditions du Muséum National D'Histoire Naturelle.

Heim, Roger, and R. Gordon Wasson. 1958. *Les Champignons hallucinogènes du Mexique: Etudes ethnologiques, taxinomiques, biologiques, physiologiques et chimiques*. With the collaboration of Albert Hofmann, Roger Cailleux, A. Cerletti, Arthur Brack, Hans Kobel, Jean Delay, Pierre Pichot, Th. Lemperière, and J. Nicolas-Charles. Paris: Editions du Muséum National d'Histoire Naturelle.

Heizer, Robert F. 1944. Mixtum compositum: the use of narcotic mushrooms by primitive peoples. *CIBA Symposia* 5: 1713-1716.

Hermans, H. G. M. 1998. *Memories of a Maverick: Andrija Puharich M.D., LL.D.* Maasslius, Netherlands: Pi Publications.

Hofmann, Albert. 1980. *LSD: My Problem Child*. New York: McGraw-Hill.

———, Roger Heim, A. Brack, and H. Kobel. 1958. Psilocybin, ein psychotroper wirkstoff aus dem mexikanischen rauschpilz *Psilocybe mexicana* Heim. *Experientia* 14: 107-109.

———, Roger Heim, A. Brack, H. Kobel, A. Frey, H. Ott, Th. Petrzilka, and F. Troxler. 1959. Psilocybin and Psilocin, zwei psychotrope Wirkstoffe aus mexikanischen Rauschpilzen. *Helvetica Chimica Acta* 42: 1557-1572.

Hoogshagen, Searle. 1959. Notes on the sacred (narcotic) mushroom from Coatlán, Oaxaca, Mexico. *Bulletin of the Oklahoma Anthropological Society* 7: 71-74.

Huxley, Aldous. 1954. *The Doors of Perception*. New York: Harper.

Johnson, Jean Bassette. 1939. The elements of Mazatec witchcraft. *Ethnological Studies* 9: 128-150.

Krieger, Louis C. C. 1967. *The Mushroom Handbook*. New York: Dover.

La Barre, Weston. 1972. Hallucinogens and the shamanic origins of religion. In *Flesh of the Gods: The Ritual Use of Hallucinogens*, ed. Peter T. Furst, pp. 261-278. New York: Praeger.

Lipp, Frank J. 1971. Ethnobotany of the Chinantec Indians, Oaxaca, Mexico. *Economic Botany* 25 234-244.

———. 1990. Mixe concepts and uses of entheogenic mushrooms. In *The Sacred Mushroom Seeker*, ed. Thomas J. Reidlinger, pp. 151-159. Portland, Oregon: Dioscorides.

Lord, Bob, and Penny Lord. 2002. *Miracles of the Child Jesus*. Morrilton, Arkansas: Journeys of Faith, Holy Family Mission.

Masters, R. E. L., and Jean Houston. 1966. *The Varieties of Psychedelic Experience*. New York: Dell.

Miller, Walter S. 1956. *Cuentos Mixes*. Instituto Nacional Indigenista, Mexico DF: Biblioteca de Folklore Indigena.

———. 1966. El *tonalamatl* mixe y los hongos sagrados. In *Summa Antropológica en Homenaje a Roberto J. Weitlaner*, ed. Antonio Pompa y Pompa, pp. 317-328. Mexico City: Instituto Nacional de Antropología e Historia, Secretaría de Educacíon Pública.

Muir, John Kenneth. 2001. *An Analytical Guide to Television's* One Step Beyond, *1959-1961*. Jefferson, North Carolina: McFarland and Company.

Neher, Andrew. 1980. *The Psychology of Transcendence*. Englewood Cliffs, New Jersey: Prentice-Hall.

Ott, Jonathon. 1976. *Hallucinogenic Plants of North America*. Berkeley, California: Wingbow.

———. 1990. A Twentieth Century Darwin. In *The Sacred Mushroom Seeker*, ed. Thomas J. Reidlinger, pp. 183-191. Portland, Oregon: Dioscorides.

———, and Jeremy Bigwood, ed. 1978. *Teonanácatl: Hallucinogenic Mushrooms of North America*. Seattle: Madrona.

Pahnke, Walter N., and William A. Richards. 1969. Implications of LSD and experimental mysticism. In *Altered States of Consciousness*, ed. Charles Tart, pp. 409-439. New York: John Wiley and Sons.

Pike, Eunice V., and Florence Cowan. 1959. Mushroom ritual versus Christianity. *Practical Anthropology* 6: 145-150.

Pollock, Steven H. 1975. The psilocybin mushroom pandemic. *Journal of Psychedelic Drugs* 7: 73-84.

Puharich, Andrija. 1959. *The Sacred Mushroom*. New York: Doubleday.

———. 1962. *Beyond Telepathy*. New York: Doubleday.

Ravicz, Robert. 1960 (1961). La mixteca en el estudio comparative del hongo alucinante. *Anales del Instituto Nacional de Antropología e Historia* 13: 73-92.

Reidlinger, Thomas J., ed. 1990. *The Sacred Mushroom Seeker: Essays for R. Gordon Wasson*. Portland, Oregon: Dioscorides.

Reyes G., Luis. 1970. Una relación sobre los hongos alucinantes. *Tlalocan* 6: 140-145.

Richardson, Allan B. 1990. Recollections of R. Gordon Wasson's 'Friend and Photographer.' In *The Sacred Mushroom Seeker*, ed. Thomas J. Reidlinger, pp. 193-203. Portland, Oregon: Dioscorides.

Rubel, Arthur J., and Jean Gettlefinger-Krejci. 1976. The use of hallucinogenic mushrooms for diagnostic purposes among some highland Chinantecs. *Economic Botany* 30: 235-248.

Ruck, Carl A. P., Blaise Daniel Staples, and Clark Heinrich. 2001. *The Apples of Apollo: Pagan and Christian Mysteries of the Eucharist*. Durham: Carolina Academic Press.

Safford, William E. 1915. An Aztec narcotic. *Journal of Heredity* 6: 291-311.

Schultes, Richard Evans. 1939. The identification of teonanacatl, a narcotic basidiomycete of the Aztecs. *Harvard Botanical Museum Leaflets* 7: 37-54.

———. 1940. Teonanacatl, the narcotic mushroom of the Aztecs. *American Anthropologist* n.s. 42: 429-443.

———. 1972. An overview of hallucinogens in the Western Hemisphere. In *Flesh of the Gods: The Ritual Use of Hallucinogens*, ed. Peter T. Furst, pp. 3-54. New York: Praeger.

Simpson, Beryl Brintnall, and Molly Conner Ogorzaly. 2001. *Economic Botany*. Third Edition, Boston: McGraw-Hill.

Singer, Rolf. 1949. The Agaricales (mushrooms) in modern taxonomy. *Lilloa* 22: 5-832.

———. 1958. Mycological investigations on Teonanácatl, the Mexican hallucinogenic mushroom, I. The history of Teonanácatl, field work and culture work. *Mycologia* 50: 239-261.

———, and Alexander H. Smith. 1958a. New species of *Psilocybe. Mycologia* 50: 141-142.

———, and Alexander H. Smith. 1958b. Mycological investigations on Teonanácatl, the Mexican hallucinogenic mushroom, II. A taxonomic monograph of *Psilocybe* Sect. *Caerulescentes. Mycologia* 50: 262-303.

Stafford, Peter. 1977. *Psychedelics Encyclopedia*. Berkeley, California: And/Or.

Stamets, Paul. 1996. *Psilocybin Mushrooms of the World*. Berkeley, California: Ten Speed.

———, and J. S. Chilton. 1983. *The Mushroom Cultivator*. Olympia, Washington: Agarikon.

Stevens, Jay. 1987. *Storming Heaven: LSD and the American Dream*. New York: Grove.

Stresser-Péan, Guy. 1990. Travels with R. Gordon Wasson in Mexico, 1956-1962. In *The Sacred Mushroom Seeker*, ed. Thomas J. Reidlinger, pp. 231-237. Portland, Oregon: Dioscorides.

Velásquez de la Cadena, Mariano, Edward Gray, and Juan L. Iribias, comp. 1957. *A New Pronouncing Dictionary of the Spanish and English Languages. With Supplement of New Words by Carlos Toral*. Chicago: Follet.

Verrill, A. E. 1914. A recent case of mushroom intoxication. *Science* n.s. 40: 408-410.

Wasson, R. Gordon. 1957. Seeking the magic mushroom. *Life*, May 13: 100-107, 109-110, 113-114, 117-118, 120.

———. 1963. The hallucinogenic mushrooms of Mexico and psilocybin: A bibliography. *Harvard Botanical Museum Leaflets* 20: 25-73.

———. 1968. *Soma: Divine Mushroom of Immortality*. New York: Harcourt, Brace & World.

———. 1973. Notes on the present status of ololiuhqui and the other hallucinogens of Mexico. In *The Psychedelic Reader*, ed. by Gunther M. Weil, Ralph Metzner and Timothy Leary, pp. 163-189. Secaucus, New Jersey: The Citadel Press.

———. 1980. *The Wondrous Mushroom: Mycolatry in Mesoamerica*. New York: McGraw-Hill.

———, George Cowan, Florence Cowan, and Willard Rhodes.1974. *María Sabina and her Mazatec Mushroom Velada*. New York: Harcourt Brace Jovanovich.

———, Stella Kramrisch, Jonathon Ott, and Carl A. P. Ruck. 1986. *Persephone's Quest: Entheogens and the Origins of Religion*. New Haven: Yale University Press.

Wasson, Valentina Pavlovna, and R. Gordon Wasson. 1957. *Mushrooms, Russia and History*. Vol. 2. New York: Pantheon.

Watts, Alan. 1962. *The Joyous Cosmology*. New York: Vintage.

Index

169

About the Author

Brian P. Akers is a mycologist and interdisciplinary scholar with additional degrees in anthropology, and comparative religion. He received his doctorate in 1997 from the Plant Biology Department of Southern Illinois University at Carbondale. Awarded for excellence in teaching at his alma mater, he has served as a professor in biology departments at several universities and colleges, including the University of Minnesota, Morris. His graduate specialization in anthropology was ethnobotany and indigenous myth, ritual, religion, shamanism, animism, witchcraft and sorcery. In addition to traditional biology classes he has developed and taught unique multidisciplinary courses in shamanism and ethnobiology which have been acclaimed by students.

As a mycologist Dr. Akers has specialized in studies of *Lepiota* s.l. (the parasol mushroom and its relatives). He is a member of the Mycological Society of America and the North American Mycological Association, has published a number of scientific papers from his research, and given talks at various institutions, and at meetings and events of nature societies and mushroom clubs. In addition to his scientific work, his broader academic pursuits have led to publications in the Bulletin of CTNS (the Center for Theology and Natural Science), and by the Metanexus Institute online.

9 780761 835820